The Economics
of Salmon Aquaculture

For
Tina Asche
and
Elizabeth Beravale

The Economics of Salmon Aquaculture

Second Edition

Frank Asche
Professor, Department of Industrial Economics, University of Stavanger, Norway

Trond Bjørndal
Professor and Director, Centre for the Economics and Management of Aquatic Resources (CEMARE), University of Portsmouth, UK, and Professor II, Aalesund University College, Norway

WILEY-BLACKWELL

A John Wiley & Sons, Ltd., Publication

Fishing News Books

First edition published 1990 © 1990 by Trond Bjørndal
This edition first published 2011 © 2011 by Frank Asche and Trond Bjørndal

Blackwell Publishing was acquired by John Wiley & Sons in February 2007. Blackwell's publishing program has been merged with Wiley's global Scientific, Technical and Medical business to form Wiley-Blackwell.

Registered Office
John Wiley & Sons Ltd, The Atrium, Southern Gate, Chichester, West Sussex, PO19 8SQ, UK

Editorial Offices
9600 Garsington Road, Oxford, OX4 2DQ, UK
The Atrium, Southern Gate, Chichester, West Sussex, PO19 8SQ, UK
2121 State Avenue, Ames, Iowa 50014-8300, USA

For details of our global editorial offices, for customer services and for information about how to apply for permission to reuse the copyright material in this book please see our website at www.wiley.com/wiley-blackwell.

Library of Congress Cataloging-in-Publication Data

Asche, Frank.
  The economics of salmon aquaculture / Frank Asche, Trond Bjørndal. – 2nd ed.
    p. cm.
  Includes bibliographical references and index.
  ISBN 978-0-8523-8289-9 (hardcover : alk. paper) 1. Salmon farming–Economic aspects. 2. Salmon–Economic aspects. I. Bjørndal, Trond. II. Title.
  SH167.S17B57 2011
  338.3'713756–dc22

                                                           2010041317

A catalogue record for this book is available from the British Library.

This book is published in the following electronic formats: ePDF (9781119993360); Wiley Online Library (9781119993384); ePub (9781119993377).

Set in 9.5/12.5pt Palatino by SPi Publisher Services, Pondicherry, India
Printed and bound in Malaysia by Vivar Printing Sdn Bhd

1 2011

Contents

Preface

The salmon aquaculture industry originated in Norway in the 1970s, and became commercially viable in the early 1980s. As a consequence of its successful development, it later spread to a number of countries in Europe, the Americas, Asia and Australia. Farmed salmon production – Atlantic salmon, coho and salmon trout – has increased from a few thousand tonnes in 1980 to about 1.9 million tonnes in 2008, and today salmon is consumed in more than 100 countries all over the world.

The Economics of Salmon Aquaculture, published in 1990, was one of the first, if not *the* first, book to systematically analyse the industry (and any aquaculture industry, for that matter) from a production and market perspective, at the firm as well as the industry level. With the tremendous development salmon aquaculture has experienced, the industry is today very different from what it was two or three decades ago. This book will bring the reader up to date on major issues pertaining to salmon aquaculture.

The book's purpose is as a textbook for senior undergraduate courses, but it should also be of use to those who work in the industry and others interested in salmon aquaculture as well as aquaculture in general. Certain parts of the book require some economic and mathematical background. However, most parts of the book are also accessible to those who lack such a background.

To a large extent, the economic analyses are based on Norwegian data. There are two reasons for this. First, as Norway is the leading salmon producer in the world, analyses of economic conditions in this country are of interest to salmon producers worldwide. Second, as Norway is also the pioneer in this field, more data are available than for any other producing country.

We would not have been able to write this book without the help and support of numerous people. Special thanks go to Sigbjørn Tveterås and Linda Nøstbakken, who have assisted with several book chapters. Christopher Martin has provided invaluable research assistance.

We also thank a number of colleagues for their help, including Jim Anderson, Atle Guttormsen, Rene Cerda, Exequiel Gonzales, Kolbjørn Giskeødegård, Jan Harald Hauvik, Gunnar Knapp, Henry Kinnucan, Daði Kristofersson, Ole Gabriel Kverneland, Laurent Le Grel, Audun Lem, Knut Molaug, Øystein Myrland, Atle Øglend, Cathy Roheim, Kristin Roll, Jan Trollvik, Ragnar Tveterås, Ursula Tveterås, Terje Vassdal and Jimmy Young. Data have been provided by Kontali (Lars Liabø, Ragnar Nystøl), the

Norwegian Seafood Export Council (Kristin Lien, Paul Aandahl, Egil Sundheim), the Norwegian Directorate of Fisheries (Merete Fauske, Per Sandberg), the Norwegian Seafood Federation (John Arne Grøttum) as well as the FAO.

The quality of the book has been enhanced by our editor, Richard Yates, while Megan Bailey has produced all the figures.

Trond Bjørndal
Frank Asche

1 Introduction

Global production of seafood – wild and farmed – has more than doubled over the past three decades, from 65 million tonnes in 1970 to 142 million tonnes in 2008 (Figure 1.1). Landings of wild fish increased from 63 million tonnes in 1970 to almost 90 million tonnes by the end of the 1980s, and have remained near that level ever since, although there are fluctuations from year to year. According to the Food and Agriculture Organisation (FAO 2009), there is little reason to expect an increase in the production from wild stocks in the foreseeable future.

During the same period, a revolution in aquaculture technology occurred, leading to a very substantial increase in production. As can be seen in Figure 1.1, aquaculture production (excluding aquatic plants) was relatively insignificant in 1970, producing about 2.5 million tonnes, or approximately 4% of total seafood production. By 2008 this had increased to 52 million tonnes, or about 37% of the total seafood supply. Consequently, even with stagnant landings of wild fish, the supply of seafood has been steadily increasing. Moreover, the supply of seafood has been growing not only in absolute quantity but also more rapidly than global population. Hence, the per-capita supply of seafood has also increased over the three last decades. The importance of aquaculture not only as a source for seafood but also for food in general is also set to continue to increase (Smith *et al.* 2010).

Salmon and shrimp are the leading species in modern industrialised aquaculture. Farmed salmonids (Atlantic and coho salmon and salmon trout) account for only about 4% of total aquaculture production but almost 13% of production value. Shrimp has a share of about 6% of production volume and accounts for more than 16% of production value. Between them, salmon and shrimp, the two most intensively farmed species, represent almost 30% of global aquaculture production value. Their production growth rates have been approximately the same as for aquaculture in total over the last 30 years. This is significant, given that they are relatively high value products. The higher value is at least partly due to the fact that these two species are among the most traded ones. In particular, most of the

The Economics of Salmon Aquaculture, Second Edition. Frank Asche and Trond Bjørndal.
© 2011 Frank Asche and Trond Bjørndal. Published 2011 by Blackwell Publishing Ltd.

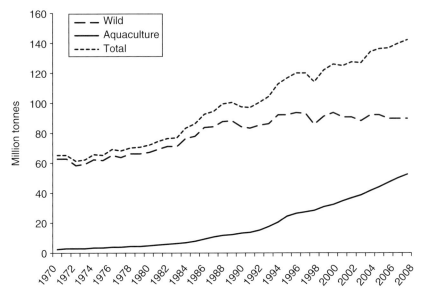

Figure 1.1   Global production of fish and seafood, 1970–2008. (*Source*: FAO)

production is consumed in the three largest markets for seafood as measured by value, the European Union (EU), Japan and the USA. Within these regions, market size was increased by expanding the geographical area where the products were consumed, as well as by reaching new consumers by reducing prices, developing new product forms and using new sales outlets.

With respect to quantity consumed, salmon is now among the top five species in most major seafood markets. Farmed salmon production – Atlantic salmon, coho and salmon trout – has increased from a few thousand tonnes in 1980 to about 1.9 million tonnes in 2008 (Figure 1.2). Salmon, together with shrimp, has in many ways led the aquaculture 'revolution' since the 1970s. While there are many differences in the development of these species, there are also a number of important similarities that other new aquaculture species are likely to emulate. Key to the aquaculture revolution is control of the production process. It is this control that makes technological innovations possible, which in turn allows lower production costs and a more affordable product to the consumers. This can be contrasted with wild harvest, where the fishers have to search for fish and have little influence over the composition of the catch. Furthermore, one has limited control over the timing of the harvest so that it is difficult to design efficient logistics systems and to address market needs.

Salmon aquaculture is of interest in itself, as well as an example with respect to opportunities and challenges that most other successful farmed species are likely to meet. It is a global industry with production on all continents except Africa, although there are two leading producing nations,

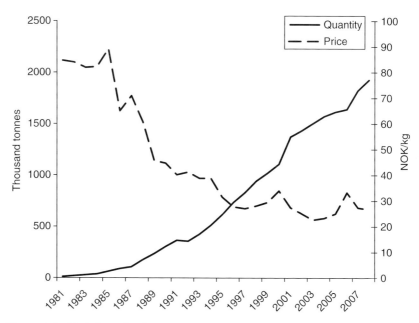

Figure 1.2   Global production of farmed salmon and real price, 1981–2008. NOK, Norwegian kroner. (*Sources*: FAO, Kontali, Norwegian Seafood Exports Council)

Chile and Norway, in two different continents and with different degrees of economic development. Salmon aquaculture is a rapidly expanding profitable industry. Moreover, it is a knowledge-based industry, and is in the forefront when it comes to technology, innovation and productivity development in aquaculture. Salmon aquaculture shows how technological innovation can create large-scale food production, and holds several important lessons with regard to how to develop aquaculture as a food production technology that increases the world's food producing ability, and as an industry capable of providing economic development in coastal communities. Finally, salmon aquaculture has experienced a number of environmental challenges that have forced farmers to change their practices to make the industry sustainable. Similar challenges are likely to face most other intensive aquaculture operations.

Control of the production process and the predictability of supply in aquaculture have profoundly changed the ways seafood can be marketed, and further contributed to the industry's success. Although control of the production process is essential for an aquaculture industry to be established, the market is equally important for a successful industry. Even with small volumes in the early 1980s, a fish buyer could achieve higher revenues or cost savings using salmon rather than other seafood because the supply was predictable. The supply could be adjusted so that more salmon were available in periods with peak demand like seasonal holidays, and also as a fresh product. As volumes grow, these advantages continue to be important. Processors like smoking houses can avoid a frozen inventory and utilise a just-in-time logistic system. Retailers can plan advertising campaigns months in advance, as with most other products (except wild seafood), and know that the quantity they require will be available when the campaign is run. The control of supply has also created the foundations for product development, as lower prices have made a new array of products affordable for consumers and profitable for processors.

The control of supply allowed sellers to target the most valuable markets, market segments and product forms. In particular, most farmed salmon is sold fresh in contrast to most wild-caught salmon that is sold frozen, tinned or in other highly conserved product forms. Fresh salmon is sufficiently valuable to justify airfreight from Europe and South America to Japan and the USA. To airfreight large volumes of seafood is but one of many innovations in logistics and distribution that control of the production process has permitted. The total result is substantial for competitiveness, as a Norwegian salmon farmer gets about 50% of the retail value of a whole salmon, as compared with 10–25% for a cod fisherman.

In the 25-year period during which salmon farming has been commercially significant, the industry, as well as the markets, has changed substantially. In particular, as shown in Figure 1.2, the strong increase in production has been associated with a substantial reduction in prices. The real price in 2008 was less than one-third of the price in the early 1980s.[1] Moreover, the price is likely to continue to decline in the longer term due to further productivity growth. The price reduction is to a large extent necessary if the industry is to continue expanding because the reduced price is the main factor in attracting new consumers of salmon.

Although we will elaborate on these issues in later chapters, they allow us to introduce the most important factors in the development of salmon aquaculture here. Control of the production process makes possible technological innovations that reduce cost of production. The reduced cost of production makes the industry profitable, and since good profits is the market's signal that it wants more of the product, this leads to increased production. However, to sell the increased production, one has to reduce prices to attract more consumers to buy salmon rather than other products. This reduces profits, and creates cycles in profitability. Over time, the equilibrium is where the produced quantity results in a price that gives the investor in the salmon industry the same risk-adjusted return on capital as in any other industry. However, with rapid innovations and market growth, this is a moving target. The cycles in profitability also create trade tensions, as most of the salmon is produced in a different country from where it is consumed.

---

[1] A value or a price expressed in real terms is adjusted for inflation to make numbers comparable over time.

In this book we will provide an analysis of what we regard as the main factors that have created the salmon aquaculture industry, as well as the main opportunities and challenges facing it. As such, the primary discussion will be both backward looking to learn from the development so far, and forward looking with respect to future development.

This book comprises two main parts. In the first part (Chapters 2–8) we focus on the development of salmon aquaculture and its lessons for aquaculture in general. In this part we discuss the biological and technical foundations for salmon aquaculture (Chapter 2), and how production has developed in the main producing countries as well as the regulations (Chapter 3). This is the foundation for discussing how productivity growth has occurred and been the main engine in the growth of salmon aquaculture (Chapter 4). A major feature in this process is how salmon aquaculture interacts with the environment, and how control of the production process allows the industry to be sustainable when externalities are taken into account (Chapter 5).

While productivity growth has been essential for the growth in salmon aquaculture, expanding the market geographically and increasing the number of product forms have also been important for the success of the industry, even though market access at times has been problematical due to trade restrictions (Chapters 6 and 7). We show that many of the characteristics of the successful development of salmon are similar to those for other successful species (Chapter 8). Hence, the insights that can be derived from salmon will to a large extent also be valid for 'new' species that are currently being developed.

In the second part (Chapters 9–11) we focus on the operation of a single firm. We start out by discussing the economic theory for optimal harvesting of the fish (Chapter 9). This model is then applied using data for western Norway to illustrate the economic decisions involved in growing and harvesting a single batch of fish (Chapter 10). This information is used as a foundation for the production plan and investment decisions for a modern farm, releasing 1 million smolts each year (Chapter 11). The investment analysis also provides the information necessary to estimate the value of an aquaculture business.

2 The Production Process in Aquaculture

Aquaculture (fish farming, fish culture, marine culture or mariculture, sea ranching) can be defined as the human cultivation of organisms in water (fresh, brackish or marine). Aquaculture is distinguished from other aquatic production by the degree of human intervention and control that is possible. It is closer in principle to forestry and animal husbandry than to traditional capture fisheries. In other words, aquaculture is stock raising rather than hunting. However, aquaculture also has a number of forms that vary with the degree of control over the production process, the species being reared, the production technology and the scale.

The production process in aquaculture is determined by biological, technological, economic and environmental factors. Many aspects of the production process can be brought under human control. The production process can be closed in the sense that it does not depend on wild stocks to provide fingerlings or fry. Environmental conditions can be controlled, breeding can be undertaken to improve yields, and harvesting timed to ensure continuous supplies of fresh product. This is in contrast to capture fisheries, which in many instances are only controlled through various harvesting regulations, if at all. And while searching for the resource is a very important part of the production process in capture fisheries, no such effort is required in aquaculture.

A number of criteria can be used to classify an aquaculture system. From an economic point of view, the most significant criterion is intensity. Commonly, production modes are divided into intensive, semi-intensive or extensive forms of culture. Measures of intensity include stocking density, production by area, feeding regimes and input costs. However, the most interesting feature is the degree of control of the production process.

In intensive aquaculture, such as salmon farming, the production system is closed. Fish are reared in confined areas and the farmer controls production factors such as farm size, stocking and feeding of fish. For salmon the confined area is a sea pen, while for other species, instead of pens, land-based

The Economics of Salmon Aquaculture, Second Edition. Frank Asche and Trond Bjørndal.
© 2011 Frank Asche and Trond Bjørndal. Published 2011 by Blackwell Publishing Ltd.

tanks (turbot), ponds (tilapia) or raceways (halibut) are used. Traditional aquaculture varies between semi-intensive and extensive. Mussel farming is an example of an extensive method, used around the globe, where the farmer provides a rope or a stake for the mussel fry to fasten, undertakes some culling so that the density does not get too high, but otherwise leaves the mussels to grow without further interference. The small ponds used in Chinese aquaculture were traditionally operated on an extensive basis, as the farmer did little to control growth and biomass. While this system is still common, many farms have become semi-intensive as farmers actively feed their fish to enhance production and undertake other productivity enhancing measures like maintaining higher densities. Recent years have seen a growing number of larger intensive facilities in China. Ocean ranching can be regarded as an extensive form of aquaculture in which a body of sea is stocked with fish that feed on natural food.

While the intensity of aquaculture production depends on the degree of control, in reality there is a continuum of operation modes. In fact, Anderson (2002) argues that the main difference between fisheries and aquaculture is the degree of control, and that the degree of control in fisheries depends on the regulatory system. He therefore argues that the continuum of production modes stretches from a high degree of control in intensive aquaculture to basically no control in unregulated fisheries. The argument is persuasive as it is sometimes hard to draw a clear distinction between aquaculture and fisheries. For instance, how much effort must an oyster fisherman put into the maintenance of his oyster beds before it becomes aquaculture?

The main focus of this book is on salmon aquaculture, which is an intensive production system. Intensive operations are necessary for aquaculture to become industrialised, and for large-scale operations. While most of the world's aquaculture production cannot be characterised as intensive today, it seems to be heading in that direction. More control over the production process allows technological innovation to a much larger extent than less intensive modes. This allows large-scale production with associated lower costs of production, a necessity if aquaculture is to fulfil its promise as a major food-producing industry that benefits the world's population at large. It also allows market-oriented production and logistics, so the fish can be sold in markets that provide the producer with higher value added.

The production process in aquaculture can be described in terms of interactions between technological and biological factors and the culture environment. This is illustrated in Figure 2.1. What is contained in each of the boxes can vary substantially for different species, production locations and markets. The physical system for a cold-water species like salmon will, for instance, differ from the physical system for a tropical species like tilapia. Similarly, the culture environment can differ substantially on the production side as well as with respect to the market where the fish is consumed.

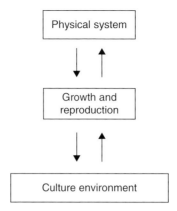

Figure 2.1   **The production system in salmon aquaculture.** (*Source*: Bjørndal (1990))

2.1   Salmon production

There are six commercially important salmon species, which all occur in nature only in the northern hemisphere. One, *Salmo salar*, is native to the Atlantic Ocean, the other five (all genus *Oncorhynchus*) to the Pacific. However, only two Pacific species, chinook and coho, are farmed.[2] In addition, in northern Europe and Chile farmed salmon trout (also genus *Oncorhynchus*, *Oncorhynchus mykiss*) are produced at a size that makes them comparable to salmon, and are primarily sold in Japan in competition with Pacific salmon.[3] This large trout is often known as salmon trout.

We will now consider the production process in aquaculture using salmon as an example, first by looking at the biological system that resembles the stages salmon go through in the wild, then at the grow-out phase and the physical system in salmon farming.

2.1.1   *Biological system*

Salmon are anadromous fish. In the wild, eggs are spawned and hatched in fresh water and the fry remain in fresh water for varying periods. Eventually, the fry go through a complex physical change known as the smoltification

---

[2]   The other Pacific salmon species are sockeye (red), chum and pink; in addition, very small quantities of cherry salmon are harvested. Some commentators also claim that a substantial part of the Alaskan and Canadian landings of wild salmon (primarily chum and pink) as well as of Japanese harvests can be regarded as aquaculture, as they are based on the release of fingerlings from hatcheries. This is known as sea ranching. However, as this is highly extensive aquaculture, we will not discuss it here.

[3]   Substantial quantities of rainbow trout and other trout varieties are farmed in various countries. However, this is primarily portion-sized fish weighing less than 0.5 kg that do not compete in what is normally thought of as the salmon market. Hence, we will not consider this production.

process. During this process they adapt to saltwater life. When this is complete, smolts migrate to sea. After spending 1–4 years at sea, depending on the species, the wild fish return to the river where they were born to spawn. After spawning, Pacific salmon always die, while Atlantic salmon can spawn more than once.

Based on the life cycle of wild salmon, the biological process in salmon aquaculture consists of the following steps:

(1) production of broodstock and roe;
(2) production of fry (hatcheries);
(3) production of smolts;
(4) production of farmed fish.

Although there are differences between Atlantic and Pacific salmon, the production cycle is essentially the same for all species.

The biological process starts with a broodstock. Originally from wild fish, broodstocks are being domesticated over time.[4] Eggs are stripped from the female, fertilised and transported to a hatchery. After an incubation period of about 2 months, yolk-sack larvae are hatched. In the wild, this takes place in January, and salmon farmers in general follow this cycle. For the first weeks, the fry feed on the contents of the yolk sack. Initial feeding, using so-called starter diets, begins after about 1 month. This is a delicate stage of the biological process and, in the early days of the industry, there was often a high mortality rate. With time and experience, the mortality rate has decreased substantially, but it still seems to be a problem for all newly farmed species. For salmon in Norway the survival rate in the hatcheries is now over 70%, which contrasts strongly with the survival rate in the wild of less than 0.5%. As the fry grow to about 5 g, they reach the fingerling phase and start developing the characteristics of salmon.

When the fingerlings become large enough, smoltification takes place, a physiological process whereby the fish are adapted to salt water. This process represents an important difference between Pacific and Atlantic salmon. Pacific salmon generally smoltify at 6–8 g, 4–6 months after being hatched. Most Atlantic salmon fry (called one-year olds) smoltify 16 months after being hatched (usually in the month of May). Year-old Atlantic salmon smolts in the wild weigh around 40 g. However, to better utilise capacity, the industry has through breeding developed smolts that develop more rapidly; 16-month-old smolts now weigh 70–140 g, a development that has allowed earlier release of smolts into the sea.

Because of the faster growth, smolts can be released to sea pens in the autumn after only 8 months. These smolts are at the lower end of the size range when released, but as they grow faster in the sea, they will be larger

---

[4] In Norway systematic breeding of salmon started in 1972; in 2008 one is using the eighth generation of this breeding stock.

than their cousins when they are released the following May. Hence, the production cycle is smoother as there are now two cohorts of salmon in the sea. Since this also makes for faster growth, it reduces the total growth time, allowing a higher turnover. As growth time is one of the most important cost factors, it is likely that autumn release will become the preferred sea transfer time in the future. However, note that the period when smolts are transferred to sea is slowly being extended. It is also worthwhile noting that the cycles are the opposite in the southern hemisphere, where Chile is the most important producing country.

Smolts are transferred to specialised grow-out farms where they are raised to marketable size in sea pens. There are substantial variations in the grow-out periods for the different species. Atlantic salmon may be raised for up to 2 years to reach weights of 2–8 kg, although normally they will not be in the sea for more than 12–18 months. Chinook will typically be raised for up to 2 years, with weights similar to those for Atlantic salmon. Coho, on the other hand, are raised for only 12–16 months to a weight of 2 kg. Depending on the firm, smolts can be produced as part of a vertically integrated operation, or purchased from independent operators.

Regardless of species, the fish must be harvested before spawning. For Atlantic and chinook salmon this occurs about 28 months after smoltification, while coho mature after only 16 months. However, it should be noted that the time of spawning can vary greatly even for fish of the same yearclass, for example a fairly high proportion of chinook and Atlantic salmon mature after 1 year. There also appear to be differences between stocks, and environmental conditions like light and water temperature are important. A large proportion of Scottish and Irish stocks tend to reach maturity as 1 year olds, while this is less common for Norwegian stocks. Although Atlantic salmon do not necessarily die after spawning, the quality degradation due to spawning would mean waiting for up to another year before harvesting. This is not practical, as the additional cost would make production unprofitable. Moreover, this would keep the pens occupied for another year, preventing the grow-out phase of a new generation. As the industry has developed, it has achieved better control over the timing of smoltification and sexual maturity, and can to a larger extent delay the processes, for example by using artificial light.

The production of fish for market can be analysed with fry or smolts as an input in the production process. This is illustrated in Figure 2.2. It is important to note that, in aquaculture, reproduction is a separate activity. In the capture fisheries, quantity harvested will determine the size of the spawning stock. This in turn will influence (future) recruitment to the stock. In salmon aquaculture, the link between stock and recruitment is broken. This allows quality-enhancing activities like breeding to take place, again making aquaculture more similar to livestock production than fishing. When an aquaculture industry has reached this stage, when it no longer depends on inputs from the wild population of the species, we say that the production process is closed.

Figure 2.2   Stock development. (*Source*: Bjørndal (1990))

2.1.2   *The grow-out phase*

The production of farmed fish in sea pens (or the **grow-out phase**) is what is normally thought of as salmon aquaculture. This is also the production phase that takes the most time, and is where most market-relevant decisions are made. As indicated in Figure 2.2, growth and natural mortality influence stock development. Both these variables are functions of time. The production of farmed fish takes a significant period of time and throughout this period capital is invested in the stock. Therefore, the time of harvesting is important (this will be analysed in detail in Chapters 9 and 10). In addition, growth is a function both of factors that the farmer can control, such as smolt quality, feeding and use of light, and of uncontrollable factors determined by the environment surrounding the farm, including seasonal variations in temperature.

Smolts are obviously a key input in the production process. Their availability, or lack thereof, can potentially limit production. However, in most years there seems to be sufficient production to make this concern irrelevant. If anything, as smolt producers would like to get all of their production sold, it is more often a concern that excess smolt production will lead to higher production of salmon and, consequently, a reduction in market prices.

Due to biological and climatic reasons, smolts can only be transferred to sea during the warmer half of the year (March to October in Norway, September to March in Chile). In nature, salmon spawn during late spring or summer, and normally hatch in January. Although smolts can in principle be transferred to sea during the summer months, due to the economics of the process May is, in practice, the latest month of significant transfer of smolts to the sea before summer. Thereafter they commence again in September. The primary consideration is making room for the next generation in the freshwater tanks used to bring salmon fry up to smolts. However, the forgone growth during a part of the best growing season (discussed in more detail in Chapters 9 and 10) also plays a role. Until the late 1990s, most smolts were released in spring, but subsequently an increased proportion of smolts have been transferred to sea during autumn. The main motivation for this is to reduce the cohort problem caused by a common release time, as all salmon are then of similar size at

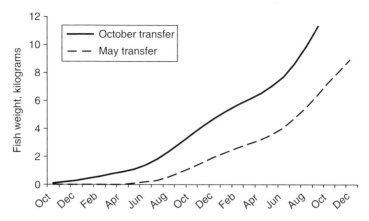

Figure 2.3   Weight curves for salmon (Norway); smolt releases in October and May.

any point in time. This is further discussed in Chapter 6, as it has given rise to some interesting price patterns, and in Chapter 10, where optimal harvesting of a yearclass of fish is considered.

The growth of the salmon can be expressed with a growth function, where growth is expressed as a function of several variables. Growth per fish over time can be represented as a function of own weight, feeding, density (the total number of fish in the pen), temperature, hours of light per day, and other biophysical factors. However, a number of studies show that all feeding regimes, other than 'feeding to saturation', will substantially increase the feed conversion ratio (i.e. feed quantity per kilogram of growth) and consequently costs (see Talbot 1993 and Einen *et al.* 1995). Hence, at any production site, there is little variation in feeding patterns when climatic and environmental variables are controlled for.[5] Accordingly, in practice, expected growth can be approximated as a function of number of days at sea at given temperatures, or day-degrees. This relationship is, of course, not exact. This means that, at any time, the farmer will have a distribution of fish of different weights around the mean predicted by this relationship.

Depending on when the smolts are transferred to sea and where the farm is located, different weight curves result. Figure 2.3 provides weight curves for a typical farm located near Bergen, Norway (approximately 70% of Norwegian fish farms are located in areas with climatic conditions similar to those around Bergen). Two curves are given: one for smolts transferred to sea water in October and the other for smolts transferred in April. The simulated growth functions show that most of the growth occurs during the warmer half of the year, late summer and early autumn.[6]

---

[5] Iwama and Tautz (1981) developed a model that simplifies the growth function by making all factors, with the exception of temperature, site-specific.

[6] It is of interest to note that wild salmon often will not feed at all during the winter months.

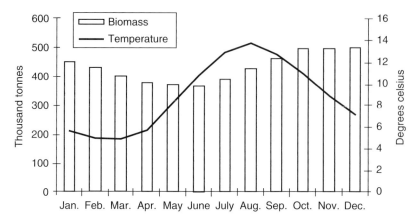

Figure 2.4 Norwegian biomass and average sea temperature, 2008. (*Source*: Norwegian Seafood Federation)

This is when the water temperature is at its highest and there is relatively more daylight (giving the highest number of day-degrees). Growth during winter is limited (the growth function stops rather abruptly, reflecting when fish become sexually mature). However, because of temperature differences, farmers in southern waters will experience higher growth than average, while farmers in northern waters will experience lower growth. Moreover, if water temperature becomes too high, growth slows down. Hence, producers in southern Norway can experience a plateau in growth in the warmest season, and in warm summers even a bimodal curve, while the curve will be more peaked for producers in a colder climate. In the UK, there has been a tendency to produce smaller fish, as the higher average water temperature gives faster growth, allowing the fish to reach marketable size before summer. The fact that salmon growth is not a linear function over time, combined with two transfers of smolts per year, gives a seasonal pattern to the availability of different weight classes of salmon for market. This also carries over to the prices, as will be shown in Chapter 6.

Average temperature and biomass development in Norway is shown in Figure 2.4. Clearly biomass development is a function of the season, although with a time lag, and there can be a substantial difference in growth between a warm and a cold year. While a growth difference of a few percent may not sound like much, the difference between a good and a bad year is easily 30 000 tonnes or more. This is enough to have an effect on price, and is an indication of how difficult it is to plan production. Growth variation can also be inferred from feeding patterns. Figure 2.5 shows feed supplies to Norwegian farmers in 2008, a year when the total feed consumption was 1 009 602 tonnes. Most of the feed was delivered during the main growing season (July to November).

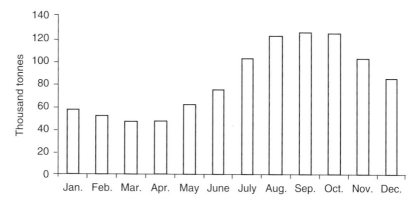

Figure 2.5   Norwegian feed deliveries by month, 2008. (*Source*: Norwegian Seafood Federation)

2.1.3   *The physical system*

In the production phase, salmon are usually raised in sea pens. In the early days of the industry single pens were used, while today most systems have up to 14 separate pens or cages. These systems were developed in the 1980s and have now become the predominant technology. They come in two main forms, steel cages and plastic cages. Plastic cages tend to be better able to handle rough weather and are cheaper, while steel cages are more robust in other ways. For instance, steel cages are better able to withstand predators like seals. Pen size has increased steadily. In the early 1980s, a single plastic cage typically had a diameter of about 5m, and reached about 4m below the surface. In 2010, a new cage may be 50m in diameter and extend 40m below the surface. With pen systems of 6–14 cages, it is clear that production capacity at each location has increased substantially.

In the early days of the industry the pen system was often attached to shore, and feed and other production equipment was stored on shore. However, as pen size has increased this is generally no longer possible. The main technology is now installations where pens are attached to a float or work barge that contains offices for the crew, control room and stores. It is possible to move the whole installation from one location to another. Moreover, with some systems, pens can be protected from heavy weather by covering them at the top and submerging them to a depth where the waves do not affect the structure. This increases the opportunity for offshore aquaculture, particularly in areas where the weather is generally good but where there are periodic violent storms.

Another technology is land-based farms, where production is undertaken in tanks or raceways with water pumped from the sea. While this is necessary for some species because of the high degree of control over production factors this technology permits, due to high investment costs it has not become common for salmon. However, it is the most usual technology for raising fry until smoltification.

Feeding methods have undergone substantial technological development, from hand feeding to automated data-controlled feeders. In the beginning, all feeding was done by hand, but high labour costs led to a substantial interest in automating this process. In the 1980s, various time-governed automatic feeders appeared, putting a given quantity of feed into the pen at specified time intervals. However, there was no way of determining whether the salmon actually ate the feed, and feeding by hand was regarded as more efficient since it led to less waste, though at a higher operating cost. The technology developed further, and today salmon farms use computerised feeding systems with sensors that detect to what extent the salmon are eating the feed. This development, in combination with feed that sinks very slowly, has reduced waste to very low levels. Hence, today automatic systems feed the fish more efficiently than hand feeding, and at a lower cost.

Other farm equipment is also becoming increasingly sophisticated to enhance growth and efficiency. Light is used to delay sexual maturity. Oxygen can be added to the pens to ensure sufficiently high oxygen levels in the water at all times. Measurement and information technology has improved substantially, giving farmers much more control over what goes on in the pens. When it comes to production facilities and monitoring the fish stock, salmon farming is increasingly becoming like any other form of modern husbandry.

Bibliography

There is a large literature describing the production process and biology for different aquaculture species. Beveridge (2004) and Leikang (2007) provide general introductions to the technical and biological aspects of aquaculture. Moksness *et al.* (2003) provide a review of other mariculture species and Leung *et al.* (2007) review species and system selection. Talbot (1993) and Einen *et al.* (1995) are studies that provide more information with respect to the feeding process in salmon aquaculture, while Iwama and Tautz (1981) provide an exposition of the growth functions. Bjørndal (1990) and Shaw and Muir (1987) present early overviews from an economic perspective, including introductions to the biology of salmon. Heen *et al.* (1993) is a collection of essays with a similar purpose. Anderson (2002) is a very illuminating study of the differences between wild fisheries and aquaculture, focusing on the opportunities and challenges control of the production process gives. Knapp *et al.* (2007) gives an overview of the influence of hatcheries in Alaska.

3 The Supply of Salmon

The global salmon supply consists of both wild and farmed salmon. As shown in Figure 3.1, supply has increased substantially during the last 25 years, from about 570 000 tonnes in 1981 to 2.65 million tonnes in 2008. In 1981, the supply was essentially wild salmon, 560 000 tonnes, while farmed production was just over 10 000 tonnes. Since then, the supply of wild salmon has grown substantially, reaching historically high levels. In the last 10 years landings have varied between 700 000 and 1 million tonnes. However, what has driven most of the growth in world salmon supply has been a tremendous increase in the farmed salmon supply that has grown to over 1.9 million tonnes in 2008. The following pages provide an overview of farmed salmon production (section 3.1), consider wild salmon (section 3.2) and discuss salmon farming regulations (section 3.3). The growth of multinational companies in salmon aquaculture will be discussed in section 3.4.

3.1 Farmed salmon production

Salmon farming is a global industry. There is production on all continents with the exception of Africa. However, most of the production is concentrated in a few regions. Currently, there are two main producers, Norway and Chile, that account for more than 77% of the total production. The UK and Canada are also important producers, each with 6–7%. Accordingly, in 2008, these four countries provided over 90% of total production.

Farmed salmon production is concentrated on three species, Atlantic and coho salmon and salmon trout (Figure 3.2), but minor quantities of other species like chinook and cherry are also produced. Atlantic salmon is the dominant species, accounting for more than 76% of output in 2008. Atlantic salmon tends to be the most profitable species, and its production share, 64% in 1985, has steadily grown relative to all other farmed salmonids. Salmon trout follows at 16.2% in 2008, down from 23.7% in 1985. Coho's share, 6.6% in 2008, is down from 11.9% in 1985. Quantities of other species

The Economics of Salmon Aquaculture, Second Edition. Frank Asche and Trond Bjørndal.
© 2011 Frank Asche and Trond Bjørndal. Published 2011 by Blackwell Publishing Ltd.

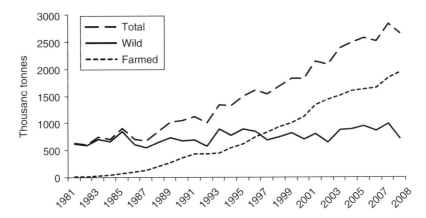

Figure 3.1   Global salmon production, 1980–2008. (*Sources*: FAO, Kontali)

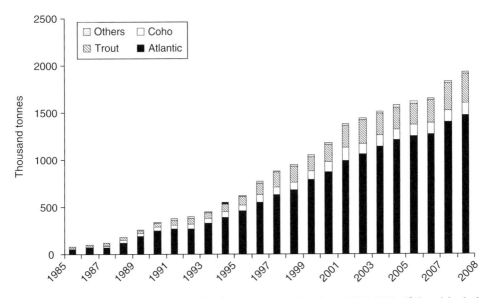

Figure 3.2   Global farmed salmon and salmon trout production, 1985–2008. 'Others' includes chinook and cherry. (*Sources*: FAO, Kontali)

remain very limited. It is of interest to note that in two of the main production regions, Chile and the west coast of Canada, Atlantic salmon is not a native species.

3.1.1   Norway

Norway's salmon farms are spread along its long coastline with its many fjords, inlets and islands that in combination with relatively stable water temperatures (ranging from 4 to 15°C) and good infrastructure provide a

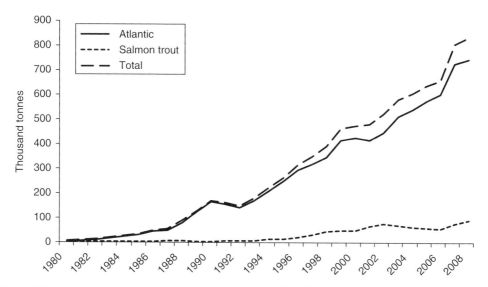

Figure 3.3   Norwegian salmon and salmon trout production, 1980–2008. (*Sources*: FAO, Kontali)

favourable environment for salmon farming. These conditions have helped make Norway the world's leading producer of farmed salmon, with an estimated output of about 827 000 tonnes round weight in 2008 (Figure 3.3). This comprised 741 000 tonnes of Atlantic salmon and 86 000 tonnes of salmon trout. From 1990 to 2008 the industry nearly quadrupled its production, with an average annual growth rate of 9.6%. Atlantic salmon production was at its highest level ever in 2008 and is still on the rise. Production has increased fairly steadily, with brief stops in 1986–1987 due to severe disease problems, and in 1990–1992 and 2001–2002 due to severe problems with profitability. Salmon trout production in 2008 was 86 000 tonnes and is still increasing. However, there have been much more marked cycles than for Atlantic salmon. The primary reason for the last cycle was a decline in production due to weaker demand in the main market for salmon trout – Japan. However, production seems set to increase again as a new market has been developed in Russia. In 2008, 222.2 million salmon smolts and 18.4 million trout smolts were released into the sea. In 2003, those numbers were 141.6 million and 16.8 million, respectively.

The tremendous growth in output, particularly throughout the 1990s, has not been matched by a corresponding increase in the number of production facilities (between 1985 and 2002 no new licences were awarded) and farm-level employment. Increased productivity in terms of feeding routines, as well as disease prevention, has improved feed conversion ratios, shortened the on-growing period and lowered mortality rates. There has also been a movement in production sites from sheltered locations, where pollution is a problem, to more exposed locations. As a result of these changes, average production costs per kilogram have dropped almost continuously since the late 1980s (this is discussed further in Chapter 4).

In 2008 there were 28 broodstock farms, 220 smolt farms and 921 salmon farm licences (on-growing sites), including salmon trout producers. The 921 salmon licences were owned by 186 companies. In total, about 4800 people were directly employed in Norwegian aquaculture, counting full-time and seasonal workers. Most of these were employed in salmon and trout production, which makes up 99% of total production. Salmon farmers constitute a highly diverse group of companies. Restrictions on ownership were lifted in 1992, and there has been an increasing tendency towards consolidation in recent years. In 2006, the four largest firms in the industry accounted for 47% of production, while the 10 largest firms accounted for about 63%. At the same time, the industry has become more international, with ownership structures across national borders. In addition to vertical integration into processing facilities, and sales offices in the EU and elsewhere, the Norwegian salmon industry has increasing ownership interests in Canadian, Chilean and Scottish salmon farming firms. A major restructuring took place in 2006 when a Norwegian company, Panfish, purchased the largest salmon-producing company, Marine Harvest, as well as another major company, Fjord Seafood.

In 2008, the Norwegian industry exported salmon worth a total of US$3.3 billion. That translated to roughly 19.8 billion Norwegian kroner (NOK), about NOK18 billion for salmon and NOK1.8 billion for salmon trout. Norway's most important market by far is the EU, although the EU share of exports has varied considerably. In 1985 it was 59%, more than 80% in 1995, and 73% in 2008. There are a number of factors influencing the variation in the EU market share, from economic and demand conditions to trade conflicts. In particular, it is likely that the numerous trade conflicts between Norway and the EU have led Norwegian exporters to target other markets. Within the EU, France is the largest market with about 16% of Norwegian export value. This makes France the largest importing country overall, followed by Poland with about 10% and Denmark with 8%. In contrast to France, Poland and Denmark are small markets with limited consumption, and most of the fish is re-exported to other EU countries.

During the last few years Russia has become the most important destination outside the EU, receiving 8% of the exports in 2008. Russia has also become the most important destination for salmon trout, with a 49% share in 2008.

For several years in the 1990s, Japan was the largest market for Norwegian salmon. However, volumes have stagnated and market share has fallen since then. Still, Japan was the second largest market outside the EU and Russia, with a 4% share in 2008. Japan has traditionally been the main market for salmon trout and had an export share of 75% in 2000. However, as exports to Russia have increased, exports to Japan have been reduced, and Japan's share in 2008 was 10%.

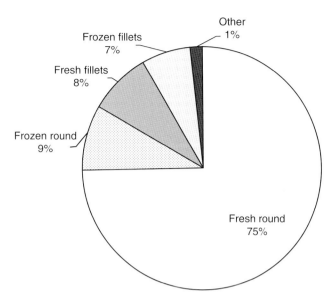

Figure 3.4 Norwegian export value by product form, 2008. (*Source*: Statistics Norway)

The USA was the largest single export market for Norwegian salmon for several years in the 1980s, with a 28.9% share in 1985. However, this market largely disappeared when anti-dumping duties were imposed in 1991.

Norway's 2008 exports are shown by product form in Figure 3.4 as a percentage of total value. Fresh round chilled salmon made up 75% of total export value. Frozen round, fresh fillets and frozen fillets comprised 9%, 8% and 7% respectively. Exports from Norway are processed in Norway only to a very limited degree. The importance of fresh fillets has been increasing in the last decade, but it is still relatively small. Other more processed product forms make up only 2% of exports. There are several reasons for this, notably high Norwegian labour costs and higher tariffs on processed products.

Outlook

Expansion during the last decade can be attributed to productivity improvements and increased output per farm, as the number of fish farm licences has remained almost unchanged since the end of the 1980s. In autumn 2002, 40 new licences were made available by the government, although only 30 were awarded. In 2003, another 50 licences were awarded, and in 2008 60 licences were advertised that were to be awarded in 2009. In addition, the introduction of a new system to measure sea pen capacity has increased production capacity at every site. As profitability has been good during the last few years, production should increase. However, trade tensions seem to be a constant barrier limiting production growth in Norway.

Norwegian salmon producers have been hard hit by trade restrictions. The countervailing duty imposed by the USA in 1991 on fresh/frozen salmon from Norway effectively eliminated this market for Norwegian producers and exporters. As a non-member of the EU, Norway faces tariffs on exports to the EU. These are considerably higher for processed (smoked, marinated, ready-to-eat) products than for unprocessed products (fresh, frozen, chilled). As a consequence, processing salmon in Norway has never become important, except for filleting. This situation is not expected to change.

Norway has also faced trade restrictions with the EU, of which there has been a series following the first dumping complaint in 1989. As a consequence of dumping and subsidy complaints by Irish and Scottish fish farmers, in 1996 the European Commission initiated an investigation of Norwegian exports. A Salmon Agreement between Norway and the EU was reached in 1997 and represented a solution to the 'Salmon Case', i.e. the investigation based on dumping allegations. The Agreement introduced a minimum price for Norwegian exports, indicative ceilings on Norwegian exports to the EU market, and a 3% marketing levy on the value of Norwegian salmon exports to the EU. Proceeds from the marketing levy were used for generic promotion of salmon in the EU. Because of the threat of trade measures, in 1995 the Norwegian government introduced a system of feed quotas for the production of salmon, limiting the amount of feed that could be used by a farm during one year. This contributed to limiting expansion in output.

The Salmon Agreement expired in March 2004 and was followed by new dumping accusations from Scottish farmers. As a consequence, the EU Commission introduced a temporary safeguard measure in the form of a quota, limiting imports from Norway, as well as other producers, as of August 2004. This was followed by further safeguards, and in 2005 by anti-dumping measures against Norwegian producers. Finally a more permanent system of minimum import prices (from June 2005) was set for 4 years in January 2006. This set of regulations was taken to the World Trade Organisation (WTO) by Norway, and following the publication of the findings of the WTO panel, the measures were abolished in 2007. After the final ruling of the WTO panel in 2008, market access seems to be less of an issue, but with continuing rapid growth there is still a danger as there will be new periods with low prices and poor profitability.

Despite regulations, trade tensions and periods with disease outbreaks, Norwegian production has been steadily increasing. From 2002 government policy has been to award a number of new licences at infrequent intervals, so that production capacity is also increasing by adding new sites. There is now more discussion in the public domain of potential negative externalities from salmon aquaculture, and competition for locations has become an issue. However, there seem to be few environmental barriers to further growth, and there are many sites that can be used by the

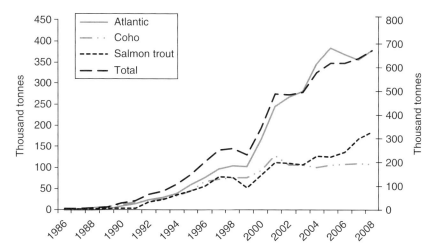

Figure 3.5   Chilean salmon production, 1986–2008. Total production on the right-hand axis. (*Sources*: FAO, Kontali)

industry if further licences are made available. Hence, Norwegian salmon production is likely to continue to grow, with output expected to exceed 1 million tonnes within a few years.

3.1.2   Chile

The Chilean salmon aquaculture industry has expanded very rapidly since the mid 1980s. Salmon are not native to Chile, but the Chilean coast provides climatic conditions very similar to their natural habitat in the northern hemisphere. In particular, good climatic conditions for salmon farming exist in the southern part of the country, where the water temperature varies even less than in Europe. The Chilean salmon industry is concentrated around Puerto Montt and Chiloé Island in Region X, about 1000 km south of Santiago, but extends also farther south into Regions XI and XII. Production focuses on Atlantic and coho salmon and salmon trout. Annual harvests can be seen in Figure 3.5. From 1119 tonnes in 1985, production increased to 673 000 tonnes in 2008, including salmon trout. Coho was initially the most important species as it was thought that a Pacific species was best suited for conditions in Chile. However, it quickly became clear that most markets preferred Atlantic salmon. By 1992 Chile was producing more Atlantic salmon than coho. Production of salmon trout accelerated in the 1990s and surpassed coho in 1997. Smaller quantities of chinook and cherry salmon have also been farmed, but are of minor importance. In 2008, Atlantic salmon made up 56.2% of production, followed by 26.6% for salmon trout and 17.2% for coho.

It is interesting to note the rapid growth in Chilean salmon production, as well as the three periods when it levelled off. From 1997 to 1999 the economic crisis in Asia hit the Chilean industry particularly hard. Moreover, this was the time when the first dumping complaint in the USA was filed against Chilean salmon. The decrease was particularly pronounced for salmon trout, where Japan was the dominant market. The next slowdown came in 2001, a period marked by weak prices in the salmon market globally. However, it is interesting to note that Atlantic salmon production continued to grow, although at a somewhat slower rate. Production growth for coho and salmon trout not only stopped but has not recovered since. A primary reason for this is that seafood demand in Japan, the main market for coho and salmon trout, has been stagnant. Finally, after an increase of about 100 000 tonnes from 2003 to 2005, production growth was again stagnant from 2005 to 2006 before it again increased until 2008. This is largely due to production problems and disease.

The Chilean government has actively promoted the development of salmon aquaculture, and regulations are more liberal here than in other salmon-producing countries. In addition to favourable environmental conditions, the industry has also benefited from low labour and feed costs, as Chile is the world's second largest producer of fish meal. The Chilean salmon industry has been developed for the most part with venture capital from large Santiago-based companies. In addition, there are no restrictions on foreign ownership in the salmon industry, and today Canadian, Japanese, Scottish and Norwegian farming interests are all represented through joint ventures or fully owned subsidiaries. The degree of concentration in the industry is fairly high, with the four largest firms accounting for 58% of production in 2006, and the 10 largest firms accounting for 85%.

The Chilean salmon industry has been geared towards export markets since the very beginning, as domestic salmon consumption is low. The main markets are the USA for Atlantic salmon, and Japan for coho and salmon trout. Most Atlantic salmon is exported fresh/chilled, but this share is declining as frozen exports increase.

The USA is the main market for fresh salmon, taking about 90% of exports. This provides the Chilean industry with an additional cost challenge, as the fish is shipped by air to the USA. Brazil has emerged as another significant buyer of fresh Atlantic salmon, taking about 10% of exports in 2008. Japan was once a relatively important market for fresh chilled salmon, but its share decreased from 11% in 1995 to only 0.3% in 2001. South American countries are becoming increasingly important markets, and although exports of whole fresh salmon have decreased substantially, it is the most important product form being shipped to Brazil and Argentina.

Frozen Atlantic salmon is primarily sent to the USA and the EU. They took 24% and 38%, respectively, of exports in 2008. However, the shares in different markets vary substantially between years, depending on market opportunities. Russia and eastern Europe are also important markets for frozen products in some years. During the last few years further-processed product for the processing and catering industries has also become increasingly

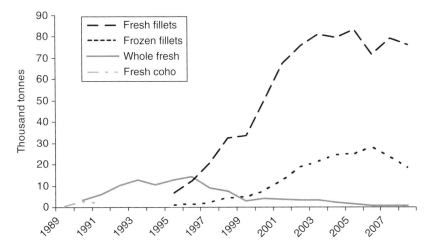

Figure 3.6   US imports of Chilean salmon, 1989–2008. (*Source*: NMFS)

important, and 24 400 tonnes of pieces and portions were exported in 2008. Again, the USA (58%) and the EU (41%) were the main markets.

For coho and salmon trout, exports are highly concentrated to the Japanese market. Both species are primarily exported frozen, and traditionally more than 90% has been shipped to Japan. However, as is the case with Norway, Russia is an increasingly important market for salmon trout, although for Chile, Japan remains the main market.

An interesting feature of the Chilean industry is that it has developed a very large export market for salmon fillets to the USA (Figure 3.6). This makes the USA an even more important market in terms of value relative to other markets. In fact, the degree of processing is higher in Chile than in Norway. This can be attributed, at least partly, to lower wages in Chile. While wage differentials have been found to have a limited effect when it comes to farming, they do appear to give Chile a competitive advantage *vis-à-vis* Norway when it comes to processing.

There is little doubt that Chile has been more market oriented than other producers. For instance, in the early 1990s, they invented the pin-bone-out fillet. Until then, the US farmed salmon market had primarily been a market along the eastern seaboard where whole salmon was presented in seafood counters. With the pin-bone-out fillets, the Chileans opened a completely new market in the Midwest, and led people who until then barely ate fish at all to consume substantial quantities.

Figure 3.6 shows Chilean exports to the USA by product form, and they clearly tell a story about market adaptation. Exports began with fresh coho, a species caught in substantial quantities by US fishermen. However, it was quickly discovered that Atlantic salmon was the preferred species along the eastern seaboard, where the main markets were located. Hence, in 1991 whole fresh Atlantic salmon took over as the leading species and product form. Fresh fillets were introduced in the early 1990s and quickly became a success, and by 1997 had taken over as the leading product form by weight. At the same time, exports of whole fresh salmon started to decline. As the market for salmon became more sophisticated, with increased processing and ready-made meals (but with discount sales at the same time), Chile also started exporting substantial quantities of frozen fillets.

Outlook

The potential for further expansion was regarded as good until 2008, and is still good in the long run. However, a serious disease outbreak resulted in a significant decline in the production of Atlantic salmon in 2009 and 2010, and also revealed several weaknesses in the regulatory structure. Hence, it may take time before the industry again reaches the production level of 2008. When the industry recovers, much of the expansion will take place in Region XI, which lacks good infrastructure. Although the industry has the potential for further cost reductions, this may to some extent be counterbalanced by greater production farther south, where cost of production is likely to be higher than in Region X.

Chile, like Norway, is very dependent on market access, and although exports to Latin America (Brazil in particular) and the EU have been increasing, Chile is still largely dependent on two markets: Japan and the USA. Chile has had its first experience with anti-dumping duties on exports to the USA, and is increasingly involved in trade issues with the EU. Moreover, vaccines are used to a lesser degree in Chile than in Europe, and diseases have become a greater concern in recent years, influencing the growth rate of the industry. Moreover, recent disease outbreaks seem more serious than earlier expected, and there are significant questions regarding what will be the Chilean production levels in the near future.

3.1.3 United Kingdom

Commercial farming of Atlantic salmon in Scotland, where all UK production takes place, commenced in the 1970s, following developments in Norway. Production expanded steadily throughout the 1980s and 1990s and totalled 160 800 tonnes in 2003. However, it declined to 120 000 tonnes in 2005 before increasing to 140 000 tonnes in 2008. The production history is shown in Figure 3.7. There are several reasons for production difficulties during the last decade. There have been problems with disease and, most importantly, profitability. Profitability was very poor during 2001–2004, making access to capital a problem for many producers. Most of the sea sites are located on the west coast and they produce 67% of the harvest, while the remaining 33% is produced on the Orkney and Shetland islands.

Marine Harvest Ltd, later a part of the Unilever Group, is generally attributed for establishing salmon farming in Scotland. As there were no

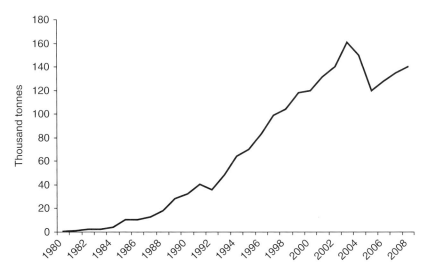

Figure 3.7   UK salmon production, 1980–2008. (*Sources*: FAO, Kontali)

restrictions on ownership in Scotland, Marine Harvest quickly became the largest salmon company in the world. The company still holds this position and has obtained a large number of locations in Chile and Norway. During the 1990s, a Dutch firm, Nutreco, owned the company. In 2006 it was taken over by a Norwegian company, Panfish, although the merged company retained the name Marine Harvest.

Fish farming has become one of the most important industries in the coastal regions of Scotland. From small beginnings the industry has grown into a multi-million pound business, employing several thousand people in some of the most remote and economically vulnerable parts of Scotland. The combined number of jobs in aquaculture and fishing is estimated at around 5000, of which about 1300 are in trout and salmon production (www.scotland.gov.uk/Publications/2008/07/11100221/17). The number of jobs in fish processing is estimated at 14 000. In addition, there are a number of jobs in services, supplies and processing.

There were 26 registered companies actively producing salmon in 2008, compared with 95 in 1998. This continues a trend of production being concentrated in fewer companies, with 95% of Scottish production now from nine companies. A further nine companies were registered as active but producing no fish for harvest in 2008. These 35 companies have 257 registered active sites (down from 343 in 1998), although only 139 sites were producing fish in 2008.

The organisation of the Scottish salmon industry underwent a radical transformation in January 2000. Concentration of power within the sector required a review of the role of the four main bodies then present, the Scottish Salmon Growers Association, the Scottish Salmon Board, Scottish Quality Salmon and the Scottish Salmon Producers Organisation (SSPO).

The organisation emerging from this review, Scottish Quality Salmon (SQS), represents an amalgamation of the first three organisations and leaves the SSPO to function as a producer organisation.

The SQS is intended to operate as a market-oriented organisation for the industry whereby the product is certified to have attained a certain standard of specifications. These standards are enforced through Food Certification Scotland and are communicated to the market via the Tartan Quality Mark through a variety of promotional instruments. The SQS is reported to account for 65% of Scottish production volume and a similar proportion of smoked output.

The emphasis of the SQS scheme is upon the quality of the Tartan Quality Mark, and this has supported some degree of product and price differentiation in the market. The most notable manifestation of this has been the award of the *Label Rouge* in France, a highly regarded recognition of quality attainment awarded to only a select range of products. Scottish salmon was the first fish and the first non-French product to achieve this status, and this has helped to ensure primacy within the French market and elsewhere.

Scotland is the only major producer of farmed salmon with a large domestic market. Nevertheless, exports are also considerable and represent roughly 50% of output. Most products are exported fresh or chilled, with continental Europe, and France in particular, as the main market. However, small volumes are sent to a number of countries to serve niche markets, and the USA is more than a niche market as it imports several thousand tonnes. Exports of cured products are also important, particularly as they are high value products.

Outlook

Potential sites for salmon farming in Scotland appear to have been exhausted. Thus, increased production will have to come from productivity improvements, unless new offshore technology should become economically viable. However, the scope for productivity improvement is substantial as Scotland has been badly affected by diseases and other production problems for a number of years.

It has proved difficult for Scottish farmers to compete with Norwegian farmers on the basis of price. The Scottish product has increasingly oriented itself to an emphasis on quality, rather than high volume with a lower price. The limitations on Scottish output, especially compared with Norway, inhibit its ability to compete on price and there seems no reason for any change in this. The emphasis on quality has permeated the supply chain, and even the more stringent health controls are benefiting quality. However, poor profitability has been a challenge, and marginal locations may face serious problems. Declining production after 2003 also indicates an industry with challenges in the future. However, production has been increasing

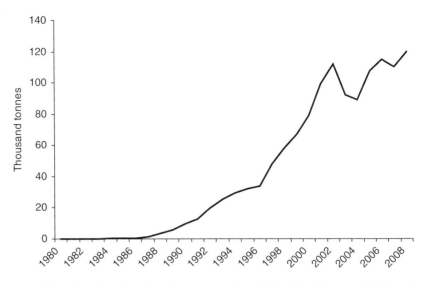

Figure 3.8   Canadian Atlantic salmon production, 1980–2008. (*Sources*: FAO, Kontali)

in recent years and is expected to again reach the level of 2003 in 2010. Hence, there seems to be potential for further growth, although limited, but competitiveness will be a challenge.

3.1.4   Canada

Salmon farming in Canada started in British Columbia (west coast) in the early 1970s and later developed in eastern Canada. Initially it focused on chinook (Canada was the largest chinook producer, with about 15 000 tonnes in 2005 but only 1400 in 2008).[7] However, as in other larger producing countries, Atlantic salmon eventually became the most important species. Production of Atlantic salmon was about 120 000 tonnes in 2008. The history of Atlantic salmon production is shown in Figure 3.8. British Columbia has about 65% and New Brunswick 30% of Canadian production.

When salmon farming developed in British Columbia, only two Pacific species, coho and chinook, were reared. Atlantic salmon was introduced at the end of the 1980s (Bjørndal 1990). Subsequently, Atlantic salmon became the preferred species, reaching 86% of British Columbia's output by volume in 2001, and has remained the dominant species ever since. No Pacific salmon is farmed in New Brunswick, Nova Scotia or Newfoundland. The Canadian industry is highly concentrated, with the four largest companies accounting for about 92% of production in 2006.

---

[7]   New Zealand has now taken over the position as the leading producer of chinook.

Canadian salmon is primarily exported in a round fresh or chilled form to the USA, with the domestic market as the second most important market. It is of interest to note that in contrast to Chile, Canadian salmon is primarily shipped as round. This indicates that, to some extent, the Canadian producers are serving different market segments than the Chilean producers.

Outlook

The scope for increased farmed salmon and salmon trout production in New Brunswick and Newfoundland is very limited for reasons of site availability and unfavourable biological conditions. There is room for some expansion in Nova Scotia and, in theory, for substantial growth in British Columbia. As a result of strong opposition to salmon farming from environmental and native groups, fishermen and residents, a moratorium on the issuance of new licences in British Columbia was imposed in 1995. The moratorium was lifted in September 2002, but given continued public opposition to the industry, it is likely that the government will proceed cautiously in granting new licences. Under the circumstances, it is difficult to predict future Canadian production. Expansion on the east coast will be limited for geographical and biological reasons. On the west coast, hostile public opinion will probably continue to hinder industry growth.

3.1.5   Other farmed salmon producers

In addition to the four main producers, there is substantial production of farmed salmon in a number of countries. We will briefly review the production in the most important ones here. However, it is worthwhile to note that with the possible exception of the Faroe Islands, these countries are likely to see their share of total production decrease, and in many cases their industries will only serve local or niche markets. It is interesting to note in Table 3.1, which shows production at 5-year intervals for five selected countries, that with the exception of the Faroe Islands production was reportedly highest in 2000 or earlier.

In the early 1980s, Japan was the second largest producer of farmed salmon in the world, primarily producing coho. Production increased steadily from 1855 tonnes in 1980 to a peak of 25 730 tonnes in 1991. Subsequently, production has been decreasing. In 2003, output was 12 000 tonnes, and in 2008 about 10 000 tonnes (Table 3.1). Because of water temperature, the growing season in Japan is fairly short; consequently, fish tend to weigh less than in Chile. Moreover, production is small scale and not as industrialised as in Chile. For these reasons, it has been difficult for Japanese farmers to compete with their Chilean counterparts.

Other producers of farmed Atlantic salmon are Ireland, the Faroe Islands, the USA (Table 3.1) and Australia and Iceland. Irish production grew to a peak of 23 700 tonnes in 2001, but has subsequently declined to 12 000 tonnes in

Table 3.1   Farmed salmon production in selected countries (thousand tonnes).

	1985	1990	1995	2000	2005	2008
USA	0	2	17	22	10	17
Ireland	2	7	12	19	12	12
Faroe Islands	1	13	8	30	17	38
Japan	9	24	16	13	12	10
Finland	7	15	17	15	13	15

Sources: FAO, Kontali.

2008. Irish output is increasingly becoming a niche product, with a strong focus on ecologically farmed salmon. The Faroe Islands increased their production substantially, to 47 000 tonnes in 2003, taking advantage of the fact that Norwegian production was limited by feed quotas. However, severe disease (particularly infectious salmon anaemia) and financial problems have stopped production at a number of sites, and in 2006 it was only 11 900 tonnes. As disease was controlled and new regulations were implemented, production again increased and reached 38 000 tonnes in 2008. Icelandic production is small and takes place in land-based facilities. The potential for increased production is limited. In Australia, where Tasmania is the centre of salmon farming, production has levelled off in recent years. In the USA, Atlantic salmon is farmed in Maine and the state of Washington. Production is not expected to increase, as new sites are not likely to become available, particularly due to environmental constraints.

Table 3.1 also shows salmon trout production in Finland. In 1985 Finland took over from Norway as the largest producer of salmon trout, and kept the leading position until 1993, when Chile supplanted it. Production peaked in 1991 at 19 100 tonnes but decreased steadily after that to a low of 12 000 tonnes in 2004. It has picked up again during the last few years. However, in contrast to Chile and Norway, Finnish production has primarily been sold in the domestic market with only limited exports. Production diminished after the domestic market was opened to competition from Norwegian salmon and salmon trout. The fact that production has picked up during the last few years is encouraging, as it indicates that the industry is becoming more competitive.

3.2   Wild salmon production

There are commercial harvests of seven wild salmon species: pink (*Oncorhynchus gorbuscha*), chum (*Oncorhynchus keta*), sockeye (*Oncorhynchus nerka*), coho (*Oncorhynchus kisutch*), chinook (*Oncorhynchus tshawytscha*),

Table 3.2   Average annual harvest (tonnes) of wild salmon by species, 1996–2005.

Species	Average 1996–2005
Chum	328 016.9
Pink	338 761.8
Sockeye	126 584.0
Coho	19 534.7
Chinook	11 771.3

Source: FAO.

masu (*Oncorhynchus masou*) and Atlantic salmon (*Salmo salar*). Commercial harvests of wild masu and Atlantic salmon are very small and are not considered further in this book. Table 3.2 gives average annual harvests for 1980–2008.

Figure 3.9 illustrates global trends in wild salmon harvests by species for 1980–2008. Total salmon harvests rose during the 1980s and early 1990s, from just over 500 000 tonnes to peaks of more than 1 million tonnes in 1995 and 2007. However, there is substantial variation in the annual landings, and they were as low as 761 000 tonnes in 2000 and 767 000 tonnes in 2008. Pink, chum and sockeye, in order, account for most commercial production. Both pink and chum harvests show an increasing trend, and account for most of the increase in total landings. This is largely due to hatcheries. They are prevalent in Alaska, as well as Japan and Canada, and primarily supply chum and pink. Sockeye landings also increased in the late 1980s and early 1990s, but have recently declined to levels similar to the early 1980s.

Figure 3.10 shows the distribution of wild salmon landings between the four main harvesting countries. The USA is the leading harvesting nation, accounting for about 43% of the total global catch over the period 1980–2008. Over the same period, Japan accounted for 28%, Russia 22% and Canada 7%. Over time, Russia has become much more important, while landings in Canada have declined.

The importance of the different Pacific salmon species differs among countries. The USA catches mainly pink and sockeye salmon, but chum has gained in relative importance in recent years, while sockeye catches have decreased. This is primarily due to increased hatchery production, and for pink salmon hatchery-based production makes up almost half of total production. Over 80% of the Japanese catches are chum, and again largely hatchery produced. Russia catches mostly pink, while Canada lands mostly pink and sockeye. It is of interest to note that Russian salmon has had a substantial impact on the Japanese market in the 1990s. This is due to the fact

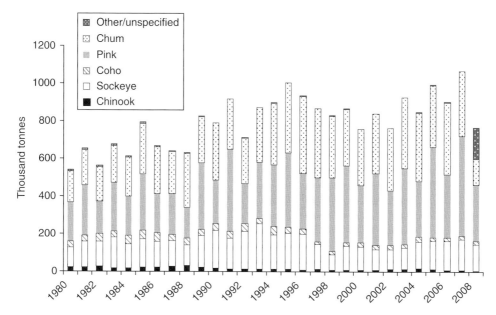

Figure 3.9 **Global wild salmon harvests by species, 1980–2008.** (*Source*: FAO)

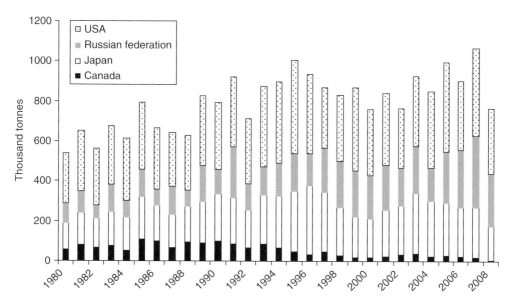

Figure 3.10 **Global wild salmon harvests by country, 1980–2008.** (*Source*: FAO)

that the old Soviet Union did not export salmon, but after the collapse of communism, Russian fishermen preferred selling to the higher paying Japanese market. Japanese vessels that buy Russian quotas also land some of the Russian salmon.

3.3   Regulation of salmon aquaculture

All the salmon-producing countries regulate their industries. Regulations are focused on ensuring that environmental standards are met and coastlines appropriately protected. In all countries an environmental impact assessment is required, although specific standards differ. As the regulatory designs differ, they also influence industry structure and competitiveness. The following provides information on regulations in Norway, Scotland and Chile. Most other producers have similar systems, although there is greater emphasis on controlling production in Norway than in other producing countries. It is also worthwhile to note that, as mentioned above, on the west coast of Canada there was a moratorium on new licences from 1995 to 2002, and environmental concerns make it very difficult for the industry to grow. In the USA environmental regulations and conflicts with respect to the use of coastal zones have essentially stopped industry development.

3.3.1   Norway

The Norwegian salmon industry has been regulated since salmon farming became commercialised, and the first set of regulations was introduced as early as 1973. From early on the impetus for establishing the industry was as a vehicle for maintaining employment in coastal communities. The first regulatory tool introduced was that a government licence was required to operate a fish farm. The law was initially applied liberally and up until 1977 all applicants were issued licences. Since then the government has controlled both the number of new licences issued and the size of farms. In addition to controlling the number of farms in the industry, there have been various requirements associated with holding a licence that have been changing over time.

The present law regulating the industry dates from 2006. Grow-out farms are regulated as to the following:

(1)   entry (licence);
(2)   location;
(3)   farm size (measured in pen volume);
(4)   ownership.

The original rationale for limiting the number of licences was to adapt production to market demand through 'balanced development' of the industry. Later, balanced development was also considered in relation to the capacity of the veterinary and extension services, education and research. This provided a further argument for a restrictive approach to awarding new licences. The Fish Farmers Association also favoured balanced development, allegedly for the same reasons.

Initially, licences were also used to influence industry structure. Until 1992, one firm (person) could hold a majority interest (51% or higher equity share) in only one fish farm. Moreover, if a majority interest in a farm was to be transferred or sold, the transaction had to be authorised by the Directorate of Fisheries. However, the extent of minority interests held by a firm (person) in fish farms was not regulated and was difficult to document. The rationale for ownership regulations was that the government wanted an owner-operator structure in the industry. The result was that most firms were single-plant operations. The exceptions were companies with multi-plant operations established prior to licensing.[8] That part of the regulations was abandoned in 1992 as it became apparent that small firms could not acquire the capital necessary to stay competitive in an industry that was becoming increasingly sophisticated. From then on restructuring has been taking place, creating larger firms that soon started to grow internationally by purchasing companies in other countries. Still there are ownership restrictions to prevent too much concentration. A special permit is required to own more than 10% of the licences, and one firm cannot own more than 20% of all licences.

During most of the industry's history, the licences were awarded for free, although often with substantial regional policy considerations. However, in 2002 the government introduced a fee of NOK5 million (except for the most northern county, Finmark, where the fee was NOK4 million). In 2002 the 10 licences (out of 40) that were not awarded were all located in Finmark, indicating that the companies thought the fee was too high for that far north (see further discussion in Chapter 11). For the licences that were advertised in 2008, the fee was increased to NOK8 million (NOK5 million for Finmark).

Ownership regulations limited firms' potential for investment in Norway. Consequently, from the early 1980s onwards, many Norwegian firms invested in fish farming abroad, particularly in Canada and the USA, but also in Chile. Thus, Norwegian capital was active in building those industries. It is clear that ownership restrictions in Norway contributed to the development of salmon farming elsewhere and thereby to increased competition from other countries.

Since 1985, broodstock farms, hatcheries and smolt producers have mainly been regulated in order to meet environmental criteria. In addition, smolt producers are regulated by ownership restrictions that aim to prevent concentration in the industry. Each operation is permitted a maximum production capacity of 2.5 million smolts annually.

Permissible production for each farm licence is regulated by pen volume. Over time, maximum size has increased from $3000\,m^3$, to $5000\,m^3$, to $8000\,m^3$ and, finally, to $12\,000\,m^3$. Regulations of this kind may prevent exploitation of economies of scale in production if efficient farm size is larger than the

---

[8]   A few firms owned more than one farm when regulations were introduced in 1973 and for that reason a few somewhat larger firms existed in Norway.

limit. The system changed in 2004, and henceforth size was limited by a restriction on maximum total biomass (MTB). As indicated by the name, the MTB sets the maximum biomass allowed at one licence. When introduced, it was set at 780 tonnes. For the farmer, an advantage with the MTB is the ability to choose the form and size of the pens without restrictions. Since the late 1990s, producers have started to combine licences, so that one production site may operate several licences.

Until 1992 the sale of salmon was regulated by a monopoly, a model associated with wild fisheries. This monopoly, known as the Fish Farmers' Sales Organisation (FOS), had stabilising salmon prices as one of its objectives. One instrument to achieve this was setting a minimum price for salmon sales from fish farms.

At the end of the 1980s and the beginning of the 1990s, there was a substantial fall in salmon prices. Consequently, profitability decreased substantially (indeed, the same was the case for wild salmon fisheries). In 1989, Scottish salmon farmers made a dumping complaint to the EU (to be further discussed in Chapter 7), and after this the FOS implemented a freezing programme for salmon to stabilise prices. The basic idea was that frozen salmon was a separate market from fresh and that Norway, with a market share of over 60% for all salmon (and even higher for fresh), would be able to exercise market power. By withholding salmon from the fresh market, it could use this market power to increase prices, or so it was thought. Unfortunately, the FOS did not have that much market power, and Figures 3.11 and 3.12 show clearly what happened. Exports of frozen salmon increased substantially in 1990–1992, but it had no effect on the price of frozen relative to fresh. Because of this freezing programme, the FOS went bankrupt in 1992. There have been no attempts to regulate prices to farmers since then. This story is worth remembering when market power and trade issues are discussed in Chapter 7.

Since the early 1990s the emphasis on adjusting production to market conditions has been reduced in Norway. However, enduring trade conflicts with the main EU market have led to several new schemes being implemented. In 1995 there was a feeding regulation to limit fish growth, and from 1996 to 2003 a system of feed quotas was in operation, where the quantity of feed for each licence was limited. This limited production, and also introduced distortions, as the firms with the most productive sites were hardest hit by the regulations. These regulations must be viewed in the context of trade restrictions in relation to the EU, and in particular the Salmon Agreement that was in operation from 1997 to 2003. This will be further discussed in Chapter 7.

3.3.2 Scotland

While there is no licensing system in Scotland as there is in Norway, and no upper limit on farm size, establishing a farm requires a lease from the Crown. In practice this works as a form of licence. Leases are issued for a

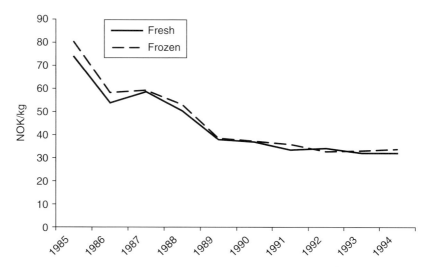

Figure 3.11   Real Norwegian export prices, 1985–1994. (*Source*: Norwegian Seafood Export Council)

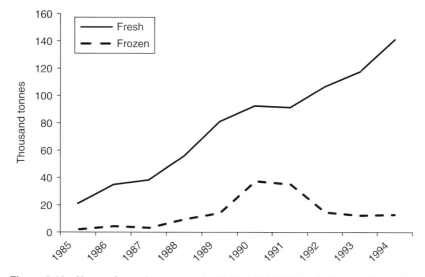

Figure 3.12   Norwegian salmon exports, 1985–1994 (1990 = 1). (*Source*: Norwegian Seafood Export Council)

10–20 year period. As the best sites have been occupied, leases have become more difficult to obtain. A system to collect economic rents was introduced as early as 1987. The rent is formula based and linked to the average selling price of salmon. The rent, which is subject to 5-year reviews, has been reduced considerably over the years, both in absolute and real terms, reflecting the decreased market value of the final product. The rent in 1987 was £48.71 per tonne, representing approximately 1% of a farm's gross turnover, while in 2003 it was set at £17.54 per tonne, about 0.8% of gross turnover (nominal values).

There are no limits on production for each lease. However, production capacity is effectively limited by environmental regulation of emissions from the site. Hence, the Scottish system uses local environmental capacity and sustainability to limit production. This also means that production capacity differs between sites, with the smallest production at shallow sites with poor flushing capacity.

There are no restrictions on ownership in the Scottish industry. The industry has therefore had a more diversified development than the Norwegian industry, with large and small firms developing side by side. Early on this allowed Marine Harvest to become the largest producer. However, it also led to investments from abroad. Marine Harvest is today a Norwegian company, and more than two-thirds of Scottish salmon production is foreign owned.

3.3.3 Chile

Salmon farming has been actively promoted by the government of Chile, to some degree as part of its plans to develop the southern regions of the country. As a consequence, Chile imposes no regulations on ownership. Licences are given for an indefinite time, and the licence owner has the right to sell it.[9] When applying for a concession (lease), a 5-year development plan must be submitted. If the concessionaire has not achieved at least 50% of the proposed activities after the first year, the concession may be reduced or retired. A modest annual fee is charged for licences.

Nevertheless, the aquaculture industry is regulated by more than 30 instruments (i.e. decrees and simple regulations) which complement the General Fisheries and Aquaculture Law in regulating various aspects of its activities, including obtaining and transferring concessions and authorisations, the importation of hydrobiological species (including eggs), the prevention and control of fish diseases, the prevention and control of environmental impacts, and the collection, processing, storage and dissemination of data on the aquaculture activity (General Fisheries and Aquaculture Law of 1991; updated 2008).

The regulatory responsibility lies with the Subsecretaria de Pesca (Undersecretariat for Fisheries), while the Servicio Nacional de Pesca – Sernapesca (National Fisheries Service) exercises the control with law compliance. Both institutions are part of the Ministry of Economy, Development and Reconstruction. Subsecretaria de Pesca is also responsible for the determination of 'Appropriate Areas for Aquaculture'. In 2003 a comprehensive

---

[9] However, it should be mentioned that an aquaculture concession and authorisation cannot be subject to any type of transfer until it has been in operation for 3 years.

National Aquaculture Policy was enacted and a National Commission of Aquaculture created. However, as a response to the disease crises in 2009, a new law has been drafted.

3.4   The growth of large multinational companies

Salmon farming, like most other aquaculture industries, started as a small-scale industry, with locally owned small companies. However, as production and marketing became more advanced, there appeared to be economies of scale and economies of scope in several of the processes. In countries where there were no regulations on ownership, relatively large companies emerged early. This first occurred in Scotland with Marine Harvest, while from early on the degree of concentration was highest in the Chilean industry. In Norway the creation of larger firms was prevented by ownership regulations. As noted in section 3.3, these regulations were not abolished until the end of 1992.

During the 1990s, there was an increased degree of concentration, and an increase in the number of multinational companies with production in several countries. Since the 1980s, Marine Harvest has been the largest company, although ownership of the company has changed several times. From its origin as a Scottish company it became the first multinational when it established a significant presence in Chile. In 1999, the owner of the company at the time, the Dutch company Nutreco, purchased Hydro Seafood, the largest Norwegian company at the time, and merged the two companies. This purchase took Marine Harvest's share of global salmon production to just above 10%. This merger was also the first time a merger in the salmon industry was investigated by competition authorities in different countries, and the enlarged Marine Harvest was not allowed to take over Hydro Seafood's Scottish operation. Hydro was the second largest producing company in Scotland after Marine Harvest, and the refusal led to the creation of a new company, Scottish Seafarmers.

In the years 2004–2006 several substantial changes took place. In 2004, when total salmon production was 1.58 million tonnes, the 10 largest companies produced about 44% of total salmonid production, as can be deduced from Table 3.3. In 2008, when total production increased to 1.94 million tonnes, their share increased to 54%, as shown in Table 3.4.

Marine Harvest was at the centre of this increase in concentration. It started with Marine Harvest merging with Stolt Sea Farm, the third largest producer in 2004. This started a spree of mergers and takeovers. The most important was Panfish of Norway's purchase of Marine Harvest and Fjord Seafood. The new company, which continues under the Marine Harvest name, is now by far the largest salmon producer in the world, even though Panfish's Scottish operation had to be sold off. This is now Scotland's second largest producer, operating under the name Lighthouse Caledonian.

Table 3.3 The world's largest salmonid producers in 2004.

Ranking	Group	Location of head office	Production (tonnes)
1	Marine Harvest	Netherlands	191 500
2	Aquachile	Chile	76 000
3	Stolt Sea Farm	Norway	74 800
4	Fjord Seafood	Norway	74 600
5	Cermaq	Norway	67 700
6	Panfish	Norway	62 200
7	Pesquera Camanchaca	Chile	43 000
8	Pesquera Los Fiordos	Chile	35 000
9	Cultivos Marinos Chiloe	Chile	35 000
10	Salmones Multiexport	Chile	34 000

Source: Kontali.

Table 3.4 The world's largest salmonid producers in 2008.

Ranking	Group	Location of head office	Production (tonnes)
1	Marine Harvest	Norway	398 300
2	Cermaq	Norway	113 700
3	Aquachile	Chile	113 500
4	Lerøy Seafood	Norway	103 000
5	Cooke Aquaculture	Canada	78 000
6	Grieg Seafood	Norway	57 500
7	Norway Royal Salmon	Norway	54 000
8	Pesquera Camanchaca	Chile	48 300
9	Pesquera Los Fiordos	Chile	46 900
10	Salmones Antartica	Japan	33 300

Source: Intrafish.

Moreover, Marine Harvest is the largest salmon producer operating in all the four largest salmon-producing countries, Canada, Chile, Norway and Scotland. In 2008, Marine Harvest's share of salmon production was about 23%. The company also had interests in several other species like halibut and sturgeon.

As can be seen by comparing Tables 3.3 and 3.4, the takeovers and mergers of Marine Harvest, Stolt Sea Farm, Fjord Seafood and Panfish substantially changed the list of the world's leading salmon producers. However, it is interesting to note that the number two and three producers in 2008 were substantially larger than they were in 2004. The main reason for this is again takeovers and mergers.

Tables 3.3 and 3.4 also show the location of the headquarters of the different companies and, as expected, Norway and Chile dominate. However, as it is a multinational industry, it is of interest to note that Marine Harvest in 2004 had its headquarters in the Netherlands. Moreover, in 2008, the 10th largest producer, Salmones Antartica, was located in Japan although most of its production was in Chile.

The location of corporate headquarters is becoming a less important indicator of ownership as the size of the companies grows. During the 1990s, an increasing number of the larger companies were listed on stock exchanges, and share ownership was international. For instance, although Marine Harvest was a part of the Nutreco group and was listed as a Dutch company, Nutreco's international ownership made the company truly international. When Marine Harvest merged with Panfish, the headquarters was moved to Norway. However, more than 80% of the shares were owned by international investors and less than 20% were Norwegian owned.

It will be difficult for Marine Harvest to grow further by takeovers and mergers as its moves are sure to be scrutinised by competition authorities. However, it is likely that there will be more mergers between other large companies in the industry. The difference in size between Marine Harvest, with about 23% of global production, and Cermaq, at about 7%, is substantial. While there will be room for some small independent companies it seems likely that a substantial part of production will be carried out by large multinational companies, due to economies of scale and the large supermarket chains' demand for efficient logistics and traceability. As such, the salmon industry has a way to go compared with the catfish industry for example, where four producer/processors account for more than 80% of supply. However, the industry seems to be following that pattern, as the four largest producers have more than 90% of production in Canada and Scotland. The degree of concentration is much lower in the two largest producing countries, as the four largest companies have about 58% of production in Chile and just under 50% in Norway. Accordingly it will be some time before the industry is dominated by a few multinational companies, if it ever happens at all.

Bibliography

The supply of salmon is reviewed in several places. FAO statistics give a good overview of total production, and how it is divided by country and production method. More details are provided by the different countries' authorities with responsibility for aquaculture or their statistical offices. Several consultants and industry organisations are very useful for more details. In addition there are several useful books, although most overviews typically are in the form of research reports. Shaw and Muir (1987) provide an early overview and a good description of the supply of wild salmon. Bjørndal (1990) focuses more on farmed salmon, while Bjørndal and Salvanes (1995) focus on the impact of Norwegian regulations. Bjørndal *et al.* (2003) provide a thorough overview of global supply and markets for salmon, wild and farmed. Knapp *et al.* (2007) provide a good general overview of the Alaskan salmon industry and also of global production with focus on wild salmon, while Valderrama and Anderson (2010) show the impact of salmon aquaculture on the fisheries sector. Anderson (2003) provides an overview of wild and aquaculture production of seafood, and illustrates how important leading aquaculture species have become.

Bjørndal and Aarland (1999) document the development of the Chilean salmon aquaculture industry. This was extended by Bjørndal (2002), who paid particular attention to industry structure and the competitiveness of the Chilean industry. More recently, industrial evolution has been analysed by Olson and Criddle (2008), while Asche *et al.* (2009a) give a tentative overview of the impact of the current disease crises in Chile.

Only recently have published papers reported supply elasticities. Asche *et al.* (2007a) and Andersen *et al.* (2008) both report a long-run supply elasticity of about 1.5. Andersen *et al.* (2008) also report the short run supply to be highly inelastic.

4

Productivity Growth and Technological Change

Salmon production has increased significantly during the previous decades. At the same time prices have declined substantially. A logical question to ask is why this has happened. For any product, profitability shapes production in that production tends to increase in a profitable industry while it is reduced if the industry loses money. The large increase in salmon production is a strong indicator that the industry has been profitable overall. The decline in salmon prices is a result of price reductions aimed at attracting new consumers and increasing consumption by current customers. However, for this to be profitable, production costs had to be substantially reduced. This indeed has been the case. Figure 4.1 shows real production cost and export price per kilogram for whole fresh salmon from Norway.

The main factor in reducing production costs is productivity growth through improved technologies and better production practices. This chapter discusses the reasons for lower production costs, focusing on Norway, where data are most widely available. As the largest producer of farmed salmon, Norway can also be considered a market leader. However, the results are relevant to other producing countries, as all producers have experienced a similar price development.

4.1 Declining costs

Figure 4.1 shows the decline in the real Norwegian production costs and export prices since 1985. When comparing production cost and export price, it is necessary to keep in mind that production costs are measured at an earlier stage in the production process than export price. Hence, they do not contain the costs involved in bringing the product to the consumer or, in this case, the costs up to the point of export. However, it is possible to compare trends and developments in production costs and prices over time. Both production costs and export prices have a clear downward

The Economics of Salmon Aquaculture, Second Edition. Frank Asche and Trond Bjørndal.
© 2011 Frank Asche and Trond Bjørndal. Published 2011 by Blackwell Publishing Ltd.

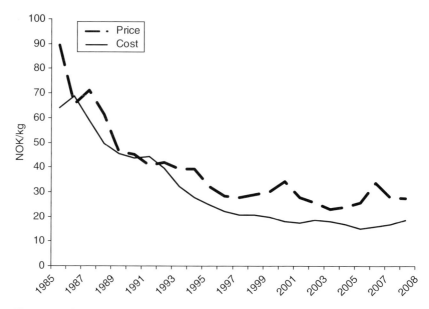

Figure 4.1   Real Norwegian production cost and export price, 1985–2008 (2008 = 1).
(*Sources*: Norwegian Directorate of Fisheries, Norwegian Seafood Exports Council)

trend, although with some obvious cycles when prices increased while production costs trended downwards. The average real export price in 2008 was 30% of the price in 1985 and production cost fell slightly more as it was only 28% of the cost in 1985. Over time, average profit margins for Norwegian farmers have remained fairly constant. Moreover, the only increases in production costs correspond to periods with diseases or special regulations and market conditions.[10] Producers and profitability were strongly affected by Hitra disease (vibriosis, caused by *Vibrio salmonicida*) in 1986 and infectious salmon anaemia and furunkulosis in 1990–1992. Further, in 1990–1992 there were special regulations for first-hand sales of salmon, and beginning in 1996 feed quotas were in effect. From 2002 to 2003 there was a slight increase in production cost, mainly attributed to poor capacity utilisation in a period with low profitability. In 2008 the international food price crisis contributed to higher costs, as feed prices increased. These issues will be further commented upon later in the chapter.

The most important insight from Figure 4.1 is that there is a close relationship between production costs and falling export prices. Hence, the factors that enable lower production costs explain a great deal of the decline in salmon prices. As production costs have declined, so have prices, keeping

---

[10]   The Directorate of Fisheries is sometimes criticised for including costs of farms affected by disease, because all salmon on a farm where a disease is found are destroyed. This leads to a statistical increase in production cost as this is counted as zero production, while the cost remains. Hence, in years when disease is a large problem, the numbers overstate the cost of production for marketed salmon.

profit margins relatively constant. This is expected in a competitive industry, as good profitability is the market's signal to increase production. As cost reductions have been translated into lower prices, it is also clear that the cost reductions have largely been passed on to consumers. The main benefit to producers is that they become larger and, accordingly, earn profits on a larger volume. This also suggests that the production cost is the main factor in determining the price. Demand will then primarily determine how much salmon is produced.

Reduced production costs are primarily due to two factors. First, fish farmers have become more efficient, as they produce more salmon with the same quantity of inputs. This is normally referred to as **productivity growth**. Second, improved input factors make the production process less costly (e.g. better feed and improved feeding technology). These show up in lower quantities of inputs used per unit of output and as lower prices for inputs. Changes in the quality of inputs or their prices can also change the mix of input factors. The distinction between productivity growth for farmers and their suppliers is often not made, and combined it is somewhat imprecisely referred to as productivity growth for the whole industry. In addition, although the focus is generally on the process of actually producing salmon, productivity growth in the distribution chain to retailers is equally important. This is because consumers are primarily interested in the final price for a product of any given quality, and whether a price reduction is due to better input factors or better logistics is of little importance.

4.2   Scale

One of the most important factors in explaining the reduction in production cost is increasing scale for the farms. Average production for each Norwegian salmon licence has increased from 47 tonnes per farm in 1982 to 904 tonnes in 2008. There are substantial variations between farms, and there are single licences with production well above 1000 tonnes due to good local conditions. Until 1992 each firm could own only one licence and production per licence and firm was the same. This remained so throughout most of the 1990s, even though firms started to merge. However, at the end of the 1990s firms that owned more than one licence started combining them. The most efficient production sites now commonly operate three to four licences, producing 3000–4500 tonnes a year. This means that the number of locations is being reduced, while the production at each location increases. In other countries there have been similar developments, although the maximum size tends to be limited by regulations, with the exception of Chile where farms traditionally have been substantially larger than in other countries.

The main reason for expanding farm size is the cost reduction that follows larger-scale operations, up to a given production level, in most industries.

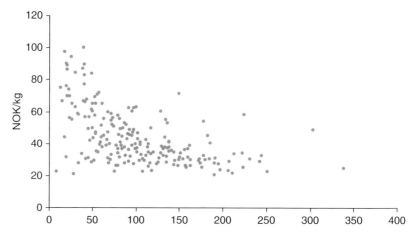

Figure 4.2 Production volume and cost in Norwegian salmon farms, 1986. (*Source*: Norwegian Directorate of Fisheries)

Economies of scale are said to be present if the unit cost of production is decreasing when quantity produced is increasing. In the early days of salmon farming there was a clear tendency for larger farms to have lower production costs, thus demonstrating economies of scale. Figure 4.2 shows the relationship between production cost and quantity produced from a survey of 206 Norwegian farms in 1986. There is a clear negative relationship between scale and cost, and farms with higher production tend to be low cost. Conversely, virtually all high-cost farms have relatively low production levels. The presence of increasing returns to scale is also seen in several econometric studies that control for other factors influencing the relationship, such as growth conditions.

Figures 4.3 and 4.4 show the same relationship for a sample of 342 farms in 1995 and 100 farms in 2004, respectively (note that scales on the axes for Figures 4.2, 4.3 and 4.4 increase over time as cost is reduced and production increased). In 1995, even the smallest farms have lower production costs and higher volume than the largest farms in 1986. The story is much the same in 2004, the only difference being a few farms with small production having higher costs. This was most likely caused by uncommon circumstances such as disease or production of special qualities. More importantly, since the mid 1990s, cost of production seems to be independent of production volume. This suggests that at this point there is no cost advantage in producing a higher volume at each farm, or that economies of scale are exhausted. However, it should be noted that this has recently been challenged when companies started to combine licences to make substantially larger units. There have been no recent econometric studies, but in the annual survey from the Norwegian Directorate of Fisheries, larger firms operating more licences now seem to have the lowest cost. Around the turn

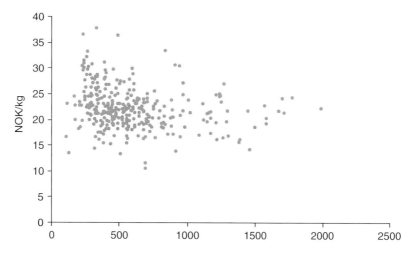

Figure 4.3   Production volume and cost in Norwegian salmon farms, 1995. (*Source*: Norwegian Directorate of Fisheries)

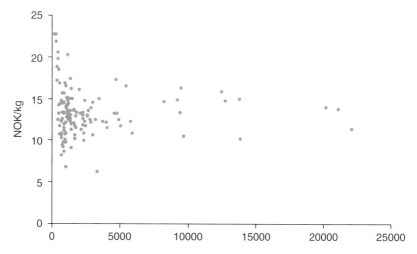

Figure 4.4   Production volume and cost in Norwegian salmon farms, 2004. (*Source*: Norwegian Directorate of Fisheries)

of the century it was the smaller firms that, on average, had the lowest cost. This, together with the fact that the new larger pen systems are being increasingly used, is an indication that there are scale economies associated with substantial increases in size. However, although the cost advantage seems to have shifted from the smallest to the somewhat larger operations, the differences are not large, and it seems that potential scale economies are relatively limited.

There is an additional factor that can be influential in determining plant size. This is the possibility for economies of scale further down the supply

Table 4.1 Nominal production cost by category in NOK per kilogram (cost shares in parentheses).

	1985	1986	1987	1988	1989	1990	1991	1992	1993	1994
Smolt cost	7.96	9.47	8.51	7.89	5.41	5.03	4.75	4.80	4.36	3.80
	(0.25)	(0.25)	(0.24)	(0.25)	(0.18)	(0.17)	(0.15)	(0.16)	(0.18)	(0.18)
Feed cost	11.00	11.45	10.50	11.25	13.01	12.92	11.69	11.94	10.49	9.84
	(0.34)	(0.3)	(0.3)	(0.36)	(0.43)	(0.43)	(0.37)	(0.41)	(0.43)	(0.47)
Insurance	1.14	1.65	1.42	1.04	1.09	1.11	0.93	0.82	0.62	0.49
	(0.04)	(0.04)	(0.04)	(0.03)	(0.04)	(0.04)	(0.03)	(0.03)	(0.03)	(0.02)
Wages	4.79	5.51	5.04	3.65	3.35	3.39	3.24	3.25	2.62	2.25
	(0.15)	(0.14)	(0.14)	(0.12)	(0.11)	(0.11)	(0.1)	(0.11)	(0.11)	(0.11)
Depreciation	1.14	1.49	1.61	1.16	1.10	1.24	1.22	1.15	0.83	0.69
	(0.04)	(0.04)	(0.05)	(0.04)	(0.04)	(0.04)	(0.04)	(0.04)	(0.03)	(0.03)
Other operating costs	3.55	4.95	4.85	3.80	3.09	3.30	6.30	4.67	3.57	2.62
	(0.11)	(0.13)	(0.14)	(0.12)	(0.10)	(0.11)	(0.2)	(0.16)	(0.15)	(0.13)
Financial cost	2.43	3.49	3.24	2.63	3.13	3.31	3.27	2.79	1.88	1.19
	(0.08)	(0.09)	(0.09)	(0.08)	(0.1)	(0.11)	(0.1)	(0.09)	(0.08)	(0.06)
Production cost	32.01	38.01	35.17	31.42	30.18	30.30	31.40	29.42	24.37	20.88

Source: Norwegian Directorate of Fisheries.

chain, for instance in slaughtering or marketing salmon. Companies indicate that control of production makes it easier to obtain these scale economies.

4.3 Structure of production costs

The cost of producing salmon has been decreasing over time. Although this can be explained in part by scale economies, at least until the mid 1990s, there are many other important factors. To gain further insight into this process it is useful to break down the cost into different components. Table 4.1 shows the average annual production cost per kilogram in nominal terms for 1985–2008, broken down into feed, smolt, capital, labour, insurance and other costs.

Feed is the most important input, accounting for 54% of operating costs in 2008. The share of feed cost has increased substantially since 1985 when it was 34%, even though the actual cost has been reduced even in nominal terms. This indicates that other inputs are used more efficiently. The cost share of smolts has been reduced from 25% in 1985 to 12% in 2008. Similarly the share for capital (depreciation and financial costs) has been reduced from 12% to 8%, labour from 15% to 9%, insurance 4% to 1%, while other costs remain steady at 12%.

Smolts are a necessary input. Once smolts have been purchased, the remaining production process is determined by the quantity of smolts.

1995	1996	1997	1998	1999	2000	2001	2002	2003	2004	2005	2006	2007	2008
3.74	3.06	2.65	2.21	2.51	2.40	2.17	2.00	1.85	1.94	1.85	1.58	2.13	2.09
(0.19)	(0.17)	(0.15)	(0.13)	(0.15)	(0.15)	(0.14)	(0.12)	(0.12)	(0.13)	(0.13)	(0.11)	(0.13)	(0.12)
9.15	8.78	8.99	9.62	8.53	7.80	7.87	9.02	8.81	8.47	7.46	8.36	9.07	9.76
(0.48)	(0.48)	(0.52)	(0.54)	(0.5)	(0.48)	(0.5)	(0.53)	(0.55)	(0.56)	(0.54)	(0.57)	(0.57)	(0.54)
0.40	0.36	0.24	0.25	0.28	0.26	0.35	0.29	0.26	0.25	0.22	0.16	0.15	0.15
(0.02)	(0.02)	(0.01)	(0.01)	(0.02)	(0.02)	(0.02)	(0.02)	(0.02)	(0.02)	(0.02)	(0.01)	(0.01)	(0.01)
1.86	1.71	1.60	1.60	1.48	1.54	1.44	1.30	1.23	1.42	1.38	1.43	1.38	1.43
(0.10)	(0.09)	(0.09)	(0.09)	(0.09)	(0.1)	(0.09)	(0.08)	(0.08)	(0.09)	(0.10)	(0.10)	(0.09)	(0.08)
0.51	0.56	0.57	0.64	0.66	0.74	0.85	0.84	0.81	0.76	0.83	0.74	0.89	0.95
(0.03)	(0.03)	(0.03)	(0.04)	(0.04)	(0.05)	(0.05)	(0.05)	(0.05)	(0.05)	(0.06)	(0.05)	(0.06)	(0.05)
2.59	2.76	2.56	2.59	2.82	2.89	2.63	2.72	1.94	1.68	1.52	2.23	1.91	2.85
(0.13)	(0.15)	(0.15)	(0.15)	(0.16)	(0.18)	(0.17)	(0.16)	(0.12)	(0.11)	(0.11)	(0.15)	(0.12)	(0.16)
0.96	0.89	0.73	0.76	0.86	0.50	0.49	0.82	1.12	0.63	0.55	0.23	0.93	0.93
(0.05)	(0.05)	(0.04)	(0.04)	(0.05)	(0.03)	(0.03)	(0.05)	(0.07)	(0.04)	(0.04)	(0.02)	(0.03)	(0.05)
19.21	18.12	17.34	17.67	17.14	16.13	15.80	17.01	16.02	15.15	13.80	14.74	15.96	18.17

However, the quality of the smolts is extremely important, as it influences survival and growth rates and the frequency of diseases. These are all factors related to breeding, which will be further discussed in section 4.7. It is important to note that variables relating to breeding influence the cost structure. In particular, in the 1980s, there was a substantial mortality rate on the farms, and perished fish could not be marketed. Hence they were a direct loss as feed and labour had been used to bring them to the point at which they perished. Further, a larger number of smolts were required to produce a given quantity of salmon. Today the survival rate is much higher (about 90%) because of better quality smolts, better husbandry practices and fewer diseases. This has contributed to reduced smolt costs. Smolts with a higher growth rate also make for better exploitation of other factors, as production is increased and turnover time reduced.

The production process has become more capital intensive as feeding and other processes have become automated. The cost share of capital is still decreasing substantially. This is because new equipment makes the process less costly to such an extent that it not only reduces labour cost, but also the cost of capital itself. Higher turnover and growth rates for salmon also better exploit capital equipment.

Labour cost has been reduced, although less than might be expected with the production process becoming more capital intensive. The use of labour has not increased to any extent since the late 1980s, even though industry output has increased from 50 000 tonnes in 1986 to 828 000 tonnes in 2008. This suggests that remuneration to labour has increased and that the skill level may have increased. Further, there has clearly been a tremendous increase in labour productivity.

The share of feed has been increasing over time, making the production process more feed intensive. As feed is the factor most closely related to production volume, this development indicates better exploitation of the capital and labour employed at each farm. This can to a large extent be explained by increased production at each farm. Several studies using data from the 1980s found that substitution was possible between feed, capital and labour. For instance, as discussed in Chapter 2, hand feeding was at the time more efficient than machine feeding. With the increased cost share of feed, these substitution possibilities have been reduced. Guttormsen (2002) suggests that substitution possibilities between input factors for a given farm to a large extent disappeared in the 1990s. This implies that salmon production now, after substantial investments in capital equipment, can be characterised as a technology with close to fixed relative factor shares in the production process. Somewhat simplified, the production process is one of converting a cheaper feed into the more desirable consumer product, salmon. When feed is produced primarily using marine inputs it becomes a process where lower value fish are converted to higher value fish. If the feed is produced using primarily vegetable inputs, these inputs are converted to fish. This also suggests that even if the substitution possibilities between capital, labour and feed are limited, farmers can substitute between different types of feed. This is indeed the case as feed can be broken down into basic proteins, and as long as one can produce feed with the correct ingredients, it does not matter where the input comes from. In current European formulas, about 35% of the feed is fish meal, which to a limited degree has been substituted for with vegetable meals, since the supply of these meals is significantly larger. About 26% of the feed is oil, of which fish oil currently makes up almost two-thirds. In Canada and Chile there is generally a lower share of marine input as differences in regulations allow a higher share of animal meals, because several retail chains in Europe require a minimum share of marine inputs in the feed.

One may think that a cost share of over 50% for feed is high. However, if one looks at comparable industries like pork and poultry, it is not very high. In particular, for the most efficient poultry producers, the cost share for feed is about 80%. This suggests that there is still a substantial efficiency potential for salmon and that production costs can be further reduced if other factors are exploited even more efficiently.

The changing composition of inputs suggests that production technology has been changing over time, and this is certainly an important factor in explaining productivity growth. Somewhat surprisingly, Tveterås and Heshmati (2002) found that technical progress at the farm level explains only about one-third of the reduction in production costs. Decreases in input factor prices, or technological innovations among the suppliers of input factors, make up the remainder. Tveterås and Heshmati (2002) also

Table 4.2   Production cost by category for smolt producers in NOK for 2008.

	Cost per smolt	Cost share
Roe and fry	0.87	(0.14)
Feeding	0.80	(0.13)
Insurance	0.08	(0.01)
Vaccination	1.16	(0.18)
Wages	1.21	(0.19)
Depreciation	0.44	(0.07)
Other operating costs	1.60	(0.25)
Net financial costs	0.22	(0.03)
Total cost	6.38	1.00

Source: Norwegian Directorate of Fisheries.

found that the time-pattern for productivity growth is anything but smooth. This is an indication that technological progress at the farm level, as well as among the suppliers, comes in leaps and is unpredictable. With the long production time in salmon farming, this can create cycles in profitability as production costs decline. This is because lower production costs initially give higher profits, inducing farmers to expand production. The expanded production then drives prices down, reducing profits.

In addition to research and innovations by the industry itself, public research and development has been important. When the industry started to show commercial potential in the mid 1970s, it was recognised that there were a number of knowledge barriers to further development. To facilitate development of an industry that could revitalise coastal communities, substantial public funds in several European countries have been invested, targeting issues of importance to the salmon industry. This was certainly an academically interesting area in its own right. However, there is little doubt that salmon farming, as a knowledge-based industry, benefited substantially from basic research aimed at a number of important issues, and that this research has accelerated the industry's development.

4.4   Smolt production

Smolts and fry are produced in locations separate from grow-out farms, and in many cases the producers are also independent. In 2008, smolts were produced on 220 sites in Norway. As one can see from Table 4.1, smolt cost has declined substantially over time, both in absolute and relative terms, and there has accordingly been a substantial productivity growth among smolt producers. This comes in addition to the quality improvement that has increased survival rates and reduced disease outbreak.

It is also of interest to look at the cost structure for smolt producers. This is provided in Table 4.2 for 2008. The most substantial difference between this cost structure and the one for grow-out operations is that there are no dominant cost factors in smolt production. Rather, the costs are relatively evenly distributed over a number of factors. It is also of interest to note that the vaccination cost is as high as 18%. This appears to be a substantial investment, although it is in general more than paid back in the form of fewer disease problems and lower use of antibiotics.

4.5   Improved feed quality

Feed is the most important input factor in the production process, and improvement in feed quality is one of the most important reasons for productivity growth. Initially, farmed fish were fed relatively crude mixes made by each farmer, based on locally landed fish and waste from nearby fish processing plants. Although farmers started to mix in fish meal and oil early on, the recipes were still crude. This type of feed is known as wet feed, and it is given to the fish in a number of forms.

The first major developments in feed quality occurred with the creation of specialised feed companies. These companies carried out systematic research on feed recipes and production techniques. A major breakthrough came in 1982, when dry feed was developed. Dry feed is primarily composed of dried (fish) meal and oil, produced as pellets. The pellets allowed large-scale production and more innovations as the formulas were improved, leading to lower costs. It also helped the development of automated feeders and subsequent feeding systems, which further contributed to productivity growth.

In the early days of the industry, unconsumed feed sank through the pens and caused major environmental problems and unnecessary costs. Moreover, to ensure fish were fed sufficiently, it was necessary to put much more feed into the pen than was actually consumed, leading to substantial waste as well as effluent emissions. Over the years, pellets have become more compact and contain more energy and materials, reducing the sinking problem. Further, feed has become slower sinking and advanced control systems have been developed so that feeding is stopped if the feed is not consumed. Hence, the feed conversion ratio or feed factor, defined as the ratio of feed to the quantity of salmon produced, has decreased over the years from 7.96 in 1988 to 1.24 in 2008 for the average farm. This has reduced emissions substantially and the quantity of wild fish needed to produce farmed salmon if the feed is based on marine fish. The feed factor is also discussed in Chapter 5.

The advent of pellets allowed new ingredients to be mixed into the feed. Some of the first additions were antibiotics to prevent or treat diseases. This is still done occasionally, but their use has been largely abandoned since the introduction of vaccines for smolts (see Chapter 5 for more discussion on antibiotic use). Another addition is astaxanthin, the substance that makes the flesh pink. Astaxanthin can be derived from shrimp, but as this is an extremely expensive process, it is primarily produced artificially. In nature, salmon get their pink colour from eating shrimp, and the colour of the flesh differs depending on their diet. By mixing astaxanthin into the feed, it is possible to control the colour of the flesh. However, it is an expensive addition and currently makes up about 10% of feed costs and 5% of total production costs.

Pellets also allow vegetable oils and meals to be added to the feed. In research facilities salmon are produced using a very low marine content in the feed. While about 5 kg of marine inputs are required to produce 1 kg of salmon using purely marine-based feed, research indicates that it is possible to produce salmon using no marine inputs at all. Today, most of the commercial feeds contain both vegetable and marine inputs, and the exact mix depends more on cost considerations than on quality. The feed formula used in Norway in 2007 contained about 35% fish meal and about 26% fish oil, requiring respectively about 2 and 2.6 kg of wild fish. As fish meal and fish oil are produced from the same species, this means that with these formulas it takes about 2.6 kg of wild fish to produce 1 kg of salmon. Hence, farmed salmon utilise feed much more efficiently than wild salmon, which require 6–8 kg of feed to grow 1 kg. The increased use of vegetable ingredients in the feed also means salmon farming does not necessarily require more marine inputs to increase production. This will primarily be a cost consideration. However, it is worthwhile to note that at very low levels of marine inputs in the feed, the growth rates for fish are significantly reduced. The levels of omega-3 fatty acids as well as taste (according to sensory panels) do not change significantly except at very low levels of marine input in the feed. Even these issues are likely to disappear as the industry obtains more knowledge about the nutritional requirements of the fish.

4.6   Diseases and increased survival rates

In nature, fish are affected by disease, and so they are in captivity. Moreover, the high densities of fish in captivity substantially increase the risk of diseases spreading. To facilitate industrial production, it must be possible to control diseases. This is certainly the case for salmon aquaculture, which throughout its history has experienced several diseases. In 1986, Norwegian farmers had the first major outbreak of vibriosis (Hitra disease), leading to substantial cost increases (see Figure 4.1). In the early 1990s, a new outbreak also caused cost increases, some due to furunkulosis and some to a new disease, infectious salmon anaemia (ISA). In the mid 1990s, a third disease appeared in Norway, infectious pancreatic necrosis (IPN), also known as 'pancreatic disease' (first discovered in Scotland in the early 1980s). At the turn of the century another disease, infectious haematopoietic necrosis (IHN), also became a problem. ISA seems to have been the most serious disease so far, and it has caused substantial problems in Canada (particularly the Bay of Fundy), the Faroe Islands and Scotland. Chile has also recently had substantial problems with ISA.

In all cases, a problem has to be discovered before it can be solved, and this is certainly true for diseases in aquaculture. In general, it is not certain which diseases will become significant problems. However, once they become problematic, cures and measures to control their spread and further outbreaks are usually found. All in all, the salmon industry appears to be relatively effective at fighting diseases. There is now a vaccination against furunkulosis, and it no longer represents a substantial problem. A commercial vaccine against IHN also exists, although there has recently been an outbreak in Canada. Trials have given positive results for a new IPN vaccine. However, ISA continues to pose a problem. The main course of action for farms infected with ISA involves quarantining the farm and destroying all fish. Provided that the distance between neighbouring farms is sufficient, this will be effective, as the infection is not easily transmitted to distant farms. However, in areas where farm density is high, a whole area can be infected, as happened in the Bay of Fundy in Canada around the turn of the century. Initially, Chile had fewer problems with disease than other producing countries, most likely because salmon are an exotic species in the region. However, major outbreaks of ISA were seen in 2007, and in 2009 the disease looks set to significantly reduce Chilean production.

In addition to these diseases a number of less serious health problems affect salmon. The most important is lice, a parasite that weakens the salmon and makes it more susceptible to other diseases. Lice are also regarded as a significant environmental problem as areas with high farmed salmon production tend to produce high levels of lice in the surrounding sea, increasing lice levels also on wild salmon.

As mentioned above, the introduction of pellets with dried feed allowed antibiotics to be included in feed, and in the mid 1980s the use of antibiotics increased to such a level that it became an environmental problem. In 1991, a smolt vaccine was introduced, leading to a quick reduction in the use of antibiotics in Norway, and today antibiotics are seldom used (see Chapter 5). The new vaccine effectively reduced health problems, and estimates are that from one generation of salmon to the next, a cost saving of NOK1–2 per kilogram was achieved in the early 1990s. However, vaccine programmes are costly and reduce growth rates, and are therefore used to different extents in producing countries.

An important problem for all farmed species is mortality in the production process, which serves to raise the farmer's total costs. In nature, salmon eggs, fries, fingerlings and finally the small fish all contribute to total production of biomass by providing feed for other species. Hence, less than 0.1% of the eggs will return as adults to spawn. In Norwegian hatcheries, the survival rate is now close to 80%, while in the grow-out phase the rates are almost 90%. This is in contrast to hatcheries for 'new' species like cod and halibut that experience survival rates of less than 10%. Although the rates in the grow-out phase are less dramatic, the survival rate has almost doubled since the early 1980s, contributing to productivity growth.

4.7   Breeding

Systematic breeding, simply described, is choosing the best parents to produce offspring with the most desirable traits. An individual acquires an equal number of genes from each parent, and they will be the principal determinants of offspring traits. The aim of systematic breeding is to use breeding animals with the best genes as parents of the next generation and at the same time avoid inbreeding.

Breeding in its simplest form is a way to help natural selection. Instead of letting nature select the strongest, breeders define some traits to be selected for. In the early 1970s, Norwegian farmers started a 4-year breeding programme for Atlantic salmon, based on fertilised eggs from Norwegian rivers. The eggs hatched in January 1972, 1973, 1974 and 1975 made up the base population of the four nucleus populations. The populations were tested in common aquaculture environments, and the fish that performed best were used as parents in the next generation. Today, four different strains of Atlantic salmon are used in one breeding programme, as the generation interval is 4 years. This means that there are 4 years between each generation of selection; however, every year in a 4-year period, the fish put into the sea are from different strains.

The initial step in a breeding programme is to select individuals with the best genetic performance for the trait of interest. The first trait selected for salmon was growth performance, measured as body weight at time of harvest. This has resulted in a near doubling of the growth rate and a reduction in the length of the production cycle to 1.5–2 years. In 1981, age at sexual maturation was added as a goal in the breeding process. As discussed in Chapter 2, the fish lose their value when they become sexually mature. Delayed sexual maturity is therefore of substantial economic value. In 1993, disease resistance was included as a breeding goal. In 1994 the first quality trait was introduced, flesh colour, followed by body composition or fat content in the meat in 1995.

Although feed utilisation is economically important, it has not been directly selected for in selection programmes due to problems with measuring feed consumption. Selection for increased growth rate has been used as an indirect selection criterion, with the assumption that this will lead to improved feed utilisation. Thodesen *et al.* (1999) compared two lines of Atlantic salmon in fresh water: offspring from salmon selected for increased growth rate for five generations (selected) and offspring from wild-caught salmon (wild) from the Namsen river, Norway. The results showed that, during the experiment, the selected line increased body weight by 79%, while the wild line increased body weight by 39%. The selected line also had significantly higher feed consumption, growth and feed utilisation. The feed efficiency ratio for the fish in the selected line was considerably higher than that for the wild line over the entire experiment. The difference indicated a 4.6% compounded increase in growth per generation of selection.

Systematic breeding programmes are of great importance in improving desirable traits in Atlantic salmon and for the growth and economic viability of the industry. In Norway, the annual cost of the breeding programme for Atlantic salmon and rainbow trout is estimated at US$3 million, paid for by the industry. In this programme, a combined individual and family selection is undertaken, and 360 families are tested each year for growth rate, age of sexual maturation, disease resistance and flesh quality. It is noteworthy that there is a trade-off between the number of goals in selective breeding and the response to each trait. The additional traits that have been introduced are leading to a reduced response to each trait. As the breeding companies have become more sophisticated and the salmon companies larger, salmon producers can also choose their own mix of traits when they engage in a long-run relationship with breeding companies.

4.8 Cycles in profitability

Movements in cost and price over time are not perfectly synchronised (see Figure 4.1). Margins between price and cost were narrow in 1986, 1991, 1997 and 2001, and wide in intervening years. Some years were much more profitable than others. This is made clear in Figure 4.5, where the unit margin (the difference between export price per kilogram for whole salmon and the production cost) is illustrated for the period 1994–2008. Although this margin cannot be interpreted as unit profit since it does not include harvesting costs (approximately NOK2.50 per kilogram), handling and transportation costs, it gives a clear impression of profitability cycles in the industry as these costs are fairly constant. Moreover, it is important to note that these are averages. There are large differences in production costs among individual companies, and although they may have similar patterns in profitability, the levels can differ substantially.

These cycles in profitability are not peculiar to salmon aquaculture.[11] They are common in other biological industries and in other industries with a substantial time lag between when the decision to increase production is made and the time the increased production enters the market. A high margin gives a signal to increase supply, but due to the time lag in

---

[11] Cycles in profitability in primary industries, and particularly in agriculture, have been studied at least since the seminal work of Ezekiel (1938).

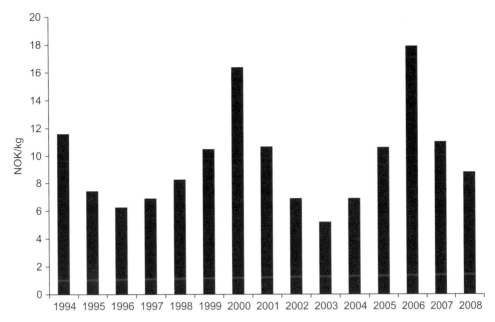

Figure 4.5 Average unit margin for Norwegian salmon farms, 1994–2008. (*Source*: authors' estimates)

production, conditions may have changed significantly when the increased output reaches the market. This often leads to over-investment, causing production to increase too much, and prices may fall for a time to, or even below, the cost of production. The low margins will then be a signal to reduce production, which again takes time, and production will often be reduced too much, giving rise to a new period with high margins.

In a stable world, producers would work out production levels that provided normal margins. However, the world is anything but stable. Production volumes that provide normal margins are a moving target because of productivity growth and other supply shocks, as well as exchange rate movements and demand shifts including market growth. The delayed responses from producers may therefore produce boom and bust cycles at irregular intervals and of varying strength.

Cycles in profitability are not a problem by themselves, as firms usually retain a substantial portion of profits at the top of a cycle to cushion the bottom of the cycle. Moreover, it requires a long-run perspective when valuing salmon firms, as it is average returns over a cycle that should matter. However, many owners do not retain earnings, causing problems at the bottom of every cycle, a feature the salmon industry shares with other primary industries. Moreover, the cycles become a problem when there are trade issues. In particular, all dumping complaints in the salmon industry can be associated with the bottom of the cycle. Trade issues will be further discussed in Chapter 7.

4.9   Catching up: regional differences

Commonly, technological progress can be divided into two parts, depending on how advanced the producers in question are. The state of the art producers will exploit the best technology, and can only improve productivity if the technical frontier is moved, i.e. the technology itself is improved. However, at any point in time there will be a number of firms that do not employ state of the art technology. These producers can improve productivity, even if the technology is not improving, by catching up with best practice. In a new industry one is more likely to observe large differences in production practices than in more mature industries, as the technical frontier is moving faster and it is more difficult to keep up with technological progress. This is certainly the case in salmon farming, as an important issue is the extent to which entrepreneurs entering the industry at later stages were able to catch up with early entrants or, more generally, how rapidly initially less efficient farmers were able to catch up with more efficient farmers.

One way to illustrate this issue is to look at regional production costs in Norway, as most of the early farms were located in the southwest before the industry spread northwards. Tveterås (1999) showed that the spread in efficiency was much larger in the mid 1980s than later. It is also the southwestern regions that are most efficient in the early part of the sample. Moreover, the least efficient regions (Finmark and Troms) experienced the highest rate of technical progress. It seems that the northern regions have been catching up with the southern regions. All regions have improved productivity, meaning that the regions that initially were least efficient have grown faster.

It also appears that convergence in the rate of technical efficiency slowed down after 1990. This may be due to different biophysical conditions (e.g. sea temperatures, tidal water exchange) or intra-regional external economies of scale in industry infrastructure, such as transportation, veterinary services and slaughtering, that give rise to permanent differences in productivity. After the late 1990s there are only minor differences in average production cost between the regions. However, there are still substantial differences between the different farms in each region.

Regional differences may comprise both productivity and technical differences. There are, of course, a number of other potential dimensions where this holds true. One is farm size. Are larger farms more efficient than smaller? Are farms that are part of larger companies more efficient because larger firms are better at spreading knowledge? The answer to both these questions seems on the whole to be no. In Norway small independent companies for a long time had the lowest production costs. This advantage has now moved to somewhat larger companies, as the most efficient firms seem to be those that operate three to four licences. This is also an indication that advantages for large companies are not found at the production level. They must be related to economies of scale, economies of scope and control in relation to downstream activities such as processing and marketing.

Of interest is productivity development in different countries. As salmon farming was pioneered in Norway, Norwegian producers would be expected to have a lead early on. However, as multinational companies were involved, producers in other countries could be expected to catch up, and this indeed seems to be the case to a large extent. This is further discussed in section 4.10.

4.10   Productivity development in Norway relative to other producers

Norwegian salmon producers have been the focus of several productivity studies. Norway is the largest producer of farmed salmon and data availability is good, while the availability of data, and therefore studies, from other producing countries is limited. Bjørndal (2002) compares cost data for Norwegian and Chilean farms and concludes that Chilean cost levels are similar to or lower than Norway's, while the cost composition is different. Chilean processing costs are lower, but transportation is higher. Chilean smolt costs are higher as survival rates are substantially lower. Industry sources normally indicate that average production costs in Scotland are €0.1–0.3 per kilogram higher than in Norway.

While lack of data prevents us from reporting on specific productivity developments among producers in different countries, it is possible to derive conclusions by investigating the evolution of production shares. In a free market, changes in production shares are caused by differences in productivity development or production costs. Improvements in productivity growth give increased market shares. In markets with trade restrictions and regulations, changes in production shares will show the combined effect of trade restrictions, regulations, and the relative productivity growth.

Figure 4.6 shows production shares for the four largest producers of all farmed salmon: Norway, Chile, the UK and Canada. Figure 4.7 provides only Atlantic salmon shares for the same producers. In 2008, these four producers combined represented about 92% of global production of farmed salmon and 94% of the production of Atlantic salmon. They have always produced more than 80% of all farmed Atlantic salmon, but their share of all farmed was as low as 60% in the mid 1980s. This was the period when Japanese production was at its highest and Finland was the leading producer of salmon trout (see Chapter 3). The figures show a very similar development, the main difference being that Norway loses relatively less share when looking at all salmon, rather than only Atlantic, as the Atlantic

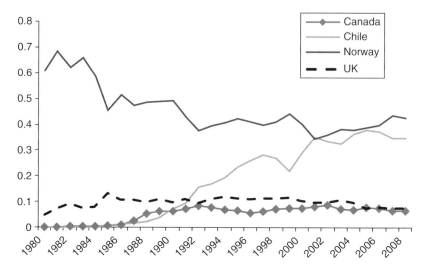

Figure 4.6 Production shares for all salmon for the four main salmon-producing countries, 1980–2008. (*Sources*: FAO, Kontali)

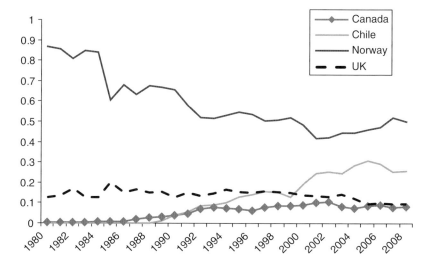

Figure 4.7 Production shares for Atlantic salmon for the four main salmon-producing countries, 1980–2008. (*Sources*: FAO, Kontali)

share of total production has increased. Moreover, Norway still clearly produces more Atlantic salmon than Chile.

The most significant trend in the figures is the evolution of Norwegian and Chilean market shares. Norway's market share fell from 69% in 1981 to 37% in 1992 for all salmon and from 85% to 52% for Atlantic salmon. Norway's share of total production reached a low of 35% in 2001. Over the same period, Chile's share grew steadily. To some extent, Norway's declining share was probably bound to happen due to diffusion of best-practice production technologies from Norway to other countries. However, there is no doubt that it was accentuated by Norwegian entry and ownership regulations, as they represented incentives to invest abroad. Since the second half of the 1980s, Norwegian capital has been invested in salmon farms in virtually all producing countries. Because of the salmon market crisis in the years 1990–1992, Norwegian ownership regulations were abandoned. A restructuring process then started as firms merged and larger firms were created, increasing Norwegian competitiveness and reducing the rate at which Norway lost market share for Atlantic salmon. And since Atlantic salmon are winning market share relative to other species, the Norwegian share for all salmon has remained stable since the early 1990s, although with some year-to-year variation. A possible reason for Norway's loss of market share for Atlantic salmon since the 1990s is new regulations introduced after anti-dumping allegations from the EU in 1996. They included feed quotas per farm that effectively limited production. However, from 2000 Norway's production share is increasing for Atlantic as well as all salmon. The steady trend of this increase indicates that productivity development in Norway is positive compared with most other salmon-producing nations.

In the 1990s, Chile became a major producer of farmed salmon. Currently, Chile is second, with about 37% of total and 29% of Atlantic production. The first major setback for Chile came in 1997–1999. It can be attributed in part to the Asian crisis of 1997–1998 that influenced demand in key markets, and in part to uncertainty following the dumping complaint in the USA. After strong growth during 1999–2001, when production in Chile equalled that of Norway, Chile lost some share between 2001 and 2003. In general, the strong positive trend for Chile is clear, and Chile has for some time looked set to surpass Norway as the largest producer in the not too distant future. The large increase in Chilean production was possible due to availability of many good locations, few restrictions on salmon farming, a low cost level, and many foreign firms in the industry providing the same knowledge base as the competitors. However, Chile also has some barriers to further growth. The first is lack of infrastructure in Region XI, where much of the future industry expansion may take place. The second is the long distance to markets, causing high transportation costs. Further, Chile's position as one of the major producers has led to anti-dumping complaints, first in the USA in 1998, and then Chile was brought into trade issues in the EU in 2002. Finally, the fact that the disease problems in Chile seem to be increasing in severity, and that production can decrease in the near future also indicates a potential for significant problems in the intermediate term.

Canada and the UK both have easy access to their main salmon markets, the USA (North American Free Trade Organisation or NAFTA) and the EU, respectively. The UK industry has been present from the early 1980s and its share peaked in 1985 with 20% of Atlantic salmon production. The Canadian industry was developing in the 1980s and increased its share then. Norwegian regulations and trade problems and Chilean trade problems were expected to benefit Canadian and Scottish salmon producers, but both countries' production shares stagnated in the 1990s. The UK share has been declining since the turn of the century. It was close to a historic low of about 9.5% of Atlantic salmon and 7.2% of all salmon in 2008, when Canada's production almost equalled that of the UK. Both the Canadian and Scottish industries seem to have experienced a productivity growth close to the industry average during most of the period, but neither producer has been able to benefit from the trade restrictions and regulations faced by Norway and Chile, and in recent years productivity growth seems to lag behind the industry average. The Canadian problem is lack of availability of sites, whereas Scotland has had disease problems and a highly valued pound sterling, in addition to more long-term constraints like availability of sites. Both reasons imply reduced profitability for Scottish farmers, and as their share has been declining the last few years, productivity development has also seemed to suffer. This is a concern for Chilean and Norwegian farmers, as it provides an incentive for anti-dumping complaints by UK producers.

The four main producers have increased their combined production share during the last decade, and their current share of more than 90% is unprecedented. The only smaller producer growing at a similar pace to the four major ones was the Faroe Islands until 2003, when there was a steep decline due to disease, but where these problems have now been overcome. On the other hand Japan, the second largest producer in the world in the early 1980s, as well as the USA, Australia, Ireland and Iceland, have fallen behind. It seems that regulations and problems with suitable locations have hindered growth to a large extent, even though production in most of these countries was for a long time growing in absolute terms. However, as discussed in Chapter 3, production has been declining in most of these countries during the last decade. It may also be that these industries, because of their small size, never realised the external scale effects associated with agglomeration and cluster effects that can be associated with a larger industry. Agglomeration effects have been identified for Norway (Tveterås, R. 2002), and most likely they are present for the other three main producing countries as well.

4.11   Cost reductions in the supply chain

Productivity growth is most easily observed in the production process and significant input factors, but another important source is improved distribution and logistics in the supply chain. As changes in the supply chain are strongly related to market development and the way salmon is marketed in different product forms, this issue will be dealt with in Chapters 6 and 7. However, when considering industry growth it is important to keep in mind that improved logistics account for a substantial part of productivity growth. Economies of scale, as well as transportation methods that have not been used for other types of fish, have reduced the cost of bringing the product to the consumer. To illustrate, consider Norwegian fresh cod exports to the UK. The fisherman's share of retail value is about 15%, while it is about 50% for salmon. If cod had the same efficient logistics, its price could be reduced by 70%. This will be further discussed in Chapters 6 and 7.

Bibliography

There are a number of papers investigating aspects of productivity growth and technological progress that has been the main engine in the production growth for salmon. Asche (1997) shows the strong relationship between the long-run trend in production cost and prices. Salvanes (1993) estimates cost functions and focuses on returns to scale and factor substitution. Østby (1999) estimates a cost function, focusing on the substitutability of capital. Tveterås (1999, 2000) estimates a production function and shows that the productivity growth also shifted the input mix, while Tveterås and Heshmati (2002) show that almost two-thirds of the productivity growth in the industry was caused by improved input factors supplied at lower prices. Guttormsen (2002) shows that the productivity growth has shifted the production technology to such an extent that in the short run (i.e. after the fish have been transferred to the sea) there is only one variable factor, feed. Asche *et al.* (2009b) show that allocative inefficiency has been reduced over time, while Nilsen (2010) shows evidence of learning by doing. Bell *et al.* (2005), Torstensen *et al.* (2005) and Turcini *et al.* (2009) studied feed ingredients and feed composition. Alfnes *et al.* (2006) and Forsberg and Guttormsen (2005, 2006) look closer at the issue of flesh colour and the cost of astaxanthin in the feed.

Bjørndal and Aarland (1999) and Bjørndal (2002) compare production costs in Norway and Chile, while Asche *et al.* (2003) examine the development of production shares. Asche *et al.* (2007b) discuss productivity improvements in the supply chain.

Bjørndal and Salvanes (1995) investigate how government regulations restrict productivity in Norwegian salmon farming. Tveterås, R. (2002) and Tveteras and Batteese (2006) show that there are industry clusters with higher productivity in Norwegian salmon farming, but that the productivity enhancing effect is reduced when the density becomes too high.

Asche and Tveterås (1999), Tveterås (1999, 2000), Kumbakhar (2001, 2002a,b) and Kumbakhar and Tveterås (2003) show that there is substantial production risk in salmon aquaculture. Moreover, production risk increases with increased scale, but the higher expected profits more than make up for the added risk. Oglend and Tveterås (2009) show that larger companies can reduce risk by operating in different regions. Asche *et al.* (2009a) and Smith *et al.* (2010) discuss the recent disease crisis in Chile.

There are also a number of papers discussing the effect of breeding. Some examples are Gjedrem (2000), Thodesen *et al.* (1999, 2001) and Kolstad *et al.* (2004). A more general overview is provided by Gjedrem and Baranski (2009).

5 Environmental Issues

Aquaculture is a production practice carried out in water and therefore impacts the environment. The increase in global aquaculture production, and the introduction of intensive farming practices in particular, has led to both local and global environmental concerns, which are relevant also for salmon. The main global concern is that increased demand for feed from a growing aquaculture production will increase fishing pressure on wild stocks and consequently threaten the sustainability of the associated capture fisheries, since marine proteins are important ingredients in the diet for cultured seafood. This is also known as the 'fish meal trap'. More local concerns include discharges from farming sites, destruction of local habitat, and spreading of pathogens. In this chapter we first discuss the fish meal trap and then local environmental concerns for salmon.

5.1 The fish meal trap

To what extent the fish meal trap represents an environmental problem depends on whether increased aquaculture production actually increases the fishing effort on species that are used for fish meal and fish oil production, also known as reduction into fish meal and fish oil. This in turn depends on two key issues. First, does the fishery management system allow increasing fish meal demand to translate into increased fishing effort and catches? If the management system functions well, this should not happen. Second, to what extent are there substitutes for fish meal, so that buyers will reduce demand for fish meal if the price increases and purchase more of alternative meals?

The aquaculture industry is not the only consumer of fish meal, even though it has grown to become the largest user, with 59% of the available quantity in 2008. As one can see from Figure 5.1, pork and poultry jointly consume 40%. For most of these species, fish meal only accounts for a small part of their diet. Other protein meals, with soya as the largest, make up the

The Economics of Salmon Aquaculture, Second Edition. Frank Asche and Trond Bjørndal.
© 2011 Frank Asche and Trond Bjørndal. Published 2011 by Blackwell Publishing Ltd.

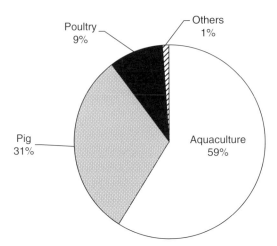

Figure 5.1 Estimated total use of fish meal, 2008. (*Source*: International Fishmeal and Fish Oil Organisation (IFFO))

major share. Fish meal production is minor compared with the total protein meal production at about 5% of total production. If fish meal is a part of the larger protein meal market, increased aquaculture production will not result primarily in increased demand for fish meal, but rather increased demand for protein meals, of which vegetable meals account for the majority. Furthermore, if fish meal is not an essential ingredient in fish feed, higher prices of fish meal will turn feed producers for aquaculture away from fish meal and thereby reduce demand from aquaculture.

5.1.1 Fisheries management

The world's reduction fisheries are mainly based on fisheries for small pelagic species.[12] Pelagic fish are used for both human consumption and reduction into fish meal and fish oil, but certain species are only fit for reduction purposes due to their consistency, often being small, bony and oily fish. Annual catches in the 1990s mainly for reduction purposes amounted to about 30 million tonnes, giving an average of 6–7 million tonnes of fish meal; in the 2000s, however, it has been close to 6 million tonnes. Production shares for fish meal in 2008 are shown in Figure 5.2. Chile and Peru combined supply 43% of global fish meal production based on their rich fisheries for Peruvian anchoveta, Chilean jack mackerel and South American pilchard. Other important producers are Thailand with

---

[12] Pelagic fish are migrating fish species that inhabit the surface waters, as opposed to demersal fish that inhabit deeper (bottom) waters.

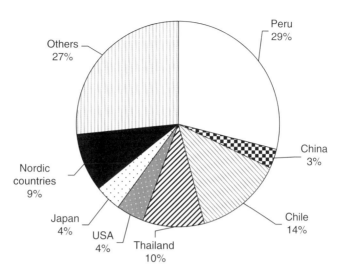

Figure 5.2   World fish meal production shares in 2008. (*Source*: IFFO)

10% and the Nordic countries Denmark, Iceland and Norway with 9% of global fish meal production.

A characteristic of pelagic fisheries is that while the quantity for human consumption stays relatively stable, the 'surplus' used for reduction can vary dramatically (Hempel 1997). Thus in years when the catches are low, such as in El Niño periods, the fish meal industry is struck hard. The pelagic fisheries have also generally been described as fully exploited or over-exploited by the FAO (Grainger & Garcia 1996). Expansion in global fish meal production, above the 6–7 million tonnes normally produced, is there-fore not likely unless prices for fish meal increase substantially. An increase in fish meal production can then happen in three ways: by increasing catches due to increased fishing effort if the management system allows it, by diverting landings that are now used for human consumption to fish meal production, or by increasing the use of cutoffs from fish processing.[13] In recent years fish meal production has been relatively stable largely because of increased use of cutoffs, and because landings of wild fish has been diverted from reduction to human consumption.

Fisheries management regimes can generally be divided into three main groups: open access, optimal management (or sole-owner management), and restricted open access. The extent to which increased demand will increase fishing pressure depends on the management regime. In the case of optimal management, the size of the landings may respond to increased prices. However, in general there is no danger of over-exploiting the stocks

---

[13]   For most species the 'fillet margin' is between 40 and 60%. A large part of the remain-der is destroyed in many countries.

since protection of a stock is a major concern in a well-managed fishery.[14] One can therefore hardly argue that the fishery poses a threat to the stock under optimal management.

In a free for all open-access fishery, the competitive behaviour of fishermen will lead to over-exploitation of fish stocks, and in some cases even to stock extinction. This situation will be exasperated by increased demand as the higher prices will lead to higher effort. This issue is particularly problematic for pelagic fish stocks, where harvesting costs are not very sensitive to stock size, so that the fishery can remain profitable even when the stock is severely over-exploited (Bjørndal 1988). Fisheries management is called for to prevent such outcomes. The simplest regulations may be to introduce catch quotas.

If a fishery is regulated by a quota that is set primarily based on biological considerations and without paying attention to economic factors, the quota remains constant when demand changes; therefore, the biomass remains the same, but the value of the catch increases. The obvious conclusion is that if the fishery is not allowed to respond to economic incentives, the increased demand for reduction species will not have much effect other than, for example, shortening season length.

Accordingly, the real problem with respect to the fish meal trap is with open access or when management is weak so that regulations are not enforced as these are the only scenarios where increased demand will lead to increased fishing pressure and increase the risk of serious stock depletion or even extinction. In reasonably well-managed fisheries, increased demand will have little impact on fishing pressure and therefore on stock size.

How then are the most important stocks used in industrial fisheries actually managed? The stocks of Peruvian anchoveta and Chilean jack mackerel have shown their vulnerability, both due to El Niño and poor fisheries management. However, fisheries management has improved over the last decade, with increasingly stricter regulations of inputs. The most important regulatory instruments used in Chile and Peru are total allowable catch (TAC) quotas, limited access, input factor regulations, and fishery closures that are imposed for certain periods and in certain areas. In Chile, individual transferable quotas have also been introduced in some fisheries. The industrial fisheries in the Nordic countries are regulated by TACs, individual vessel quotas as well as additional restrictions. The overall state of reduction fisheries in the Nordic countries has improved substantially after pelagic stock collapses in the late 1960s and early 1970s, and several of the stocks have been rebuilt to pre-collapse levels. In the USA, the menhaden fishery is the main industrial fishery, and is also regulated with TACs.

Open access is not an appropriate description of the management situation for most important pelagic species. However, quotas tend to be high

---

[14] Bjørndal (1988) showed that for stocks with sufficiently low growth rates, it may be optimal to harvest the stock to extinction. However, this point is primarily of academic interest.

and one may often question whether the sustainability of the fish stocks has been the main criterion when the quotas are set. Therefore, whether increased demand for fish meal from a growing aquaculture industry will lead to unsustainable fisheries will also to some extent depend on the market structure for fish meal.

5.1.2   *The markets for oil meals*

As shown in Figure 5.1, substantial quantities of fish meal also go to livestock production. Furthermore, as discussed above, other feed meals can also be used in aquaculture feed. There are two main reasons why fish meal is used in livestock as well as aquaculture production. One explanation stresses the uniqueness of fish meal. Fish meal has both a higher protein content and a different nutritional structure than the other protein meals. In particular, this is the case with respect to amino acids that may be positive for growth and the general health of the animals. If fish meal is unique, increased demand from aquaculture production for fish meal is likely to increase prices, and therefore increase fishing pressure on poorly managed fish stocks. Additional demand from aquaculture then has to be met partly at the expense of other users who find it too expensive, and partly because the higher price leads to increased production. The other explanation emphasises that fish meal in general is a cheap protein. If fish meal is used primarily because it is a cheap protein, one would expect a high degree of substitutability between fish meal and other protein meals. This substitution can take two forms. Either other buyers of fish meal can substitute to other meals because they find that fish meal is too expensive, or aquaculture producers can substitute away from feed with too much marine content as this becomes too expensive.

These two explanations have very different implications for the price formation process for fish meal and the impact of higher aquaculture production. If fish meal is used because it is unique, the price of fish meal should be determined by the demand for and supply of fish meal alone. If, in addition, fish meal is essential in the aquaculture feed, increased demand from aquaculture can drive demand for fish meal. However, if fish meal is a close substitute for other protein meals, one would not expect the price of fish meal to be much influenced by increased demand from aquaculture, since the price is determined by total demand for protein meals. If so, increased demand from aquaculture is not a threat to wild stocks.

'What is the market?' is then an important question since the extent of the market determines whether increased demand from aquaculture will affect prices. To determine fish meal's position in the protein meal market, Asche and Tveterås (2004) investigated its relationship with soya meal, the most important of the vegetable-based meals. Relationships between European and US-produced meals were analysed on a monthly basis for the period from January 1981 to April 1999. The soya meal and fish meal

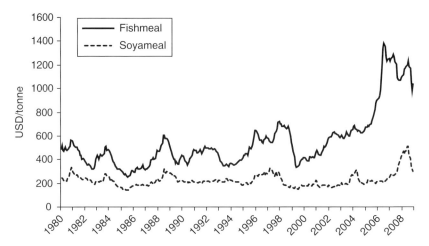

Figure 5.3 Monthly fish meal and soya meal price data from Hamburg, 1980–2008. (*Source*: FAO)

prices are shown in Figure 5.3, extended to cover the period through 2008. Note that the fish meal prices are substantially higher than the soya meal prices. This is primarily because of the higher protein content. If one adjusts for the protein content, most of the difference disappears. The period which the data span is interesting for at least two reasons. Firstly, there were some extreme situations for fish meal production in this period due to low raw material supply, including El Niños in 1982–1983, 1986–1988, 1991–1992 and 1997–1998, with the first and the last being the most severe. This makes it interesting to compare how the fish meal and soya meal markets have interacted during these extreme periods. Secondly, this is the period when most of the growth in intensive aquaculture has taken place. If fish meal is primarily demanded due to its special attributes, this should identify fish meal and soya meal as different market segments during this period.

The results of Asche and Tveterås (2004) suggest that fish meal and soya meal are strong substitutes. Therefore, until 1999, it was the total demand for fish meal and soya meal, possibly together with the demand for other protein meals, that determined the price of these protein meals. In order for aquaculture to influence the price of fish meal with this market structure, the changes in demand or supply must be large enough to affect demand and supply for fish and soya meal combined. This is important, since with this market structure it is unlikely that increased demand for fish meal from the aquaculture sector will lead to increased prices for fish meal, as it has only a negligible share of this combined market. With this market structure it seems unlikely that increased demand for fish meal from the aquaculture sector will increase fishing pressure in industrial fisheries.

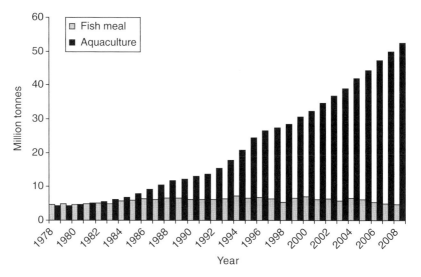

Figure 5.4   Annual aquaculture and fish meal production, 1978–2008. (*Sources*: FAO, IFFO)

However, Figure 5.3 also indicates that the relationship becomes weaker after the turn of the century, and Kristofersson and Anderson (2006) show that a structural break has taken place. In 2006, one can also observe that the fish meal price sky-rocketed relative to soya meal, indicating that fish meal to a larger extent became demanded primarily because of its unique attributes, and accordingly is no longer a part of the larger protein meals market. However, the soya meal price increased with a 1-year lag and the prices were largely realigned in 2008.

It seems like growth in aquaculture production is fairly independent of the cycles in the fish meal price and of the availability of fish meal. In Figure 5.4 we show global production of fish meal and aquaculture production. As one can see, there is a very strong growth in total aquaculture production, as has been the case for salmon production. Hence, the cycles in fish meal price do not seem to have a strong impact on aquaculture, and most aquaculture producers do not appear to require fish meal in large quantities. Hence, for most aquaculture species, fish meal does not seem to be an essential ingredient, and even when fish meal is demanded primarily for its unique attributes, aquaculture does not seem to be the main force in the increased demand. However, to the extent that there are species that require sufficient quantities of fish meal, producers of these species will find that feed costs will become more volatile, and if the price continues to increase, it may also be a problem for the profitability of the operation.

We have not commented on the other main marine input used in fish feed, fish oil. This is because this product has not received much attention as it is much easier to create oils with a similar structure to vegetabile oils. Hence, despite the fact that aquaculture is using a higher share of the world's fish oil production than of the fish meal production, a potential lack of fish oil does not appear to be an issue.

5.1.3 Concluding remarks on the fish meal trap

Increased aquaculture production has the potential to influence wild fish stocks via increased demand for feed. For this to happen, the management system for the species in question must be weak and, second, there cannot be close substitutes for fish meal either from aquaculture or from other buyers of fish meal. The most important fish stocks that are used in fish meal production are managed with quota systems, but one can question if these are sufficiently comprehensive to prevent the fish meal trap. Whether fish meal is demanded from aquaculture because of its unique properties is less clear. Fish meal has been a part of the much larger protein meal market and, in particular, fish meal is a close substitute for soya meal. With this market structure it is total supply of, and demand for, protein meal, of which fish meal makes up only 4%, that determines prices for fish meal. However, in recent years it appears that the link between fish meal and the larger protein meal market has become weaker, and accordingly that fish meal to a larger extent is demanded because of its unique attributes. On the other hand, the growth in aquaculture (and salmon) production has not been much influenced by fish meal prices, and one seems to be able (at least partly) to substitute fish meal in the feed when prices are high. One is then led to the conclusion that increased demand for fish meal from aquaculture cannot have any significant impact on fish meal prices in the long run (allowing for potential sticky adjustment in consumption) and, accordingly, does not lead to increased fishing pressure.

It is a fact that demand for fish meal from aquaculture has grown from virtually nothing to more than half of total production in only 20 years. If demand for fish meal from the aquaculture sector continues to grow, it is possible that this structure may change. However, this does not have to be the case, since it is not clear that the demand for fish meal from the aquaculture sector is due mainly to the unique characteristics of fish meal. Moreover, as we saw in Chapter 4, productivity growth leading to lower cost of production has been a major factor in the growth of salmon aquaculture, and in Chapter 8 we will show that this is the case also for other species. If the market structure changes so that fish meal is demanded because of its unique characteristics, production cost will also increase and therefore limit the demand for feed from aquaculture. Hence, even with a structural change in the fish meal market, the increase in demand from aquaculture and feed prices are likely to be limited. However, this change in production cost may well influence which species are to be large-volume species in aquaculture, as species that use feed with limited marine content will be more competitive from a production cost perspective.

Some commentators also indicate that lack of fish meal will eventually prevent further growth in aquaculture production. For this to be the case, fish meal must be an essential input in aquaculture feed. As this does not seem to be the case for most species, this concern seems unjustified.

5.2 Local issues

Any production process that interacts with the natural environment has the potential to damage the environment around the production site. This includes destruction of natural habitat and pollution from production that influences habitat and wildlife around the site. For salmon farming the main issues have been pollution from organic waste and the interaction between wild and farmed salmon. Farmed salmon may transmit diseases to wild salmon, and escaped farmed salmon may attempt to spawn in rivers and thereby impact the genetic pool for wild salmon. This experience is not unique for salmon farming. Shrimp farming has received negative publicity due to detrimental environmental effects, such as destruction of mangroves, saltification of agricultural areas, eutrophication and disruptive socioeconomic impacts. However, while poor production practices can be unsustainable, sustainability is not an issue in well-managed farms and industries.

The environmental issues that arose in intensive salmon and shrimp farming during the 1980s and 1990s must be seen in relation to the introduction of a new technology that uses the environment as an input. The larger the production at any site and the more intensive the process, the larger the potential for environmental damage. However, the greater the degree of control of the production process in intensive aquaculture also makes it easier to address these issues. With all new technologies there will be unexpected side effects, and there will be a time lag from when an issue arises until it can be addressed. First, the impact and the causes must be properly identified. Second, the solution to the problems will require modifications of existing technology or maybe entirely new technology. In both cases pollution reduction implies some form of induced innovation. In this relation, Tveterås, S. (2002) argues that industry growth can have a positive effect on pollution, in line with the environmental Kuznets curve (EKC). The EKC hypothesis refers to an empirical observation that pollution tends to increase with economic growth up to a certain level, after which growth will reduce pollution (Arrow *et al*. 1995). This gives the pollution profile over time the shape of an inverted U.

There are two main causes for the industry to address environmental effects: the effects reduce productivity and therefore profits, and/or government regulations force the industry to do so. Detrimental environmental effects of aquaculture not accounted for in market prices are by definition negative externalities. Asche *et al*. (1999a) argued that internalisation of the externalities explains why some of the major environmental issues have been resolved in aquaculture. The argument goes as follows. Productivity in aquaculture depends on an environment where farmed fish thrive. Fish farms with environmental practices that cause the quality of the local environment to deteriorate will experience negative feedback effects, where poor water quality reduces on-farm productivity. These negative environmental feedback effects are well known, and can be amply exemplified by reference to salmon and shrimp farming. The results are reduced biomass growth through deteriorating fish health and, in the worst case, disease outbreaks that wipe out entire on-farm fish stocks. Consequently, one is concerned with developing management practices that avoid such negative repercussions on productivity.

However, if there is no negative feedback on profitability, it is unlikely that the industry will internalise detrimental environmental effects. In this case the government has to regulate the industry if the effects are to be avoided. The rapid growth of global aquaculture has represented an environmental challenge for authorities. First, knowledge about the environmental effects has been limited, or at worst lacking. This has called for extensive research to identify causes and effects. Second, in many places local governments do not have the resources to implement and enforce regulations. Finally, it is desirable that regulations are efficient in addressing the externalities but, conversely, also allow the aquaculture industry to be economically sustainable if that is possible.

We will look closer at the main local environmental issues in relation to Norwegian salmon aquaculture, namely emissions due to organic waste, the use of antibiotics and chemicals, and escapees and sea lice. We look at Norway as this is the only region where we have data.

5.2.1 Organic waste

Effluent discharges are one of the major environmental concerns in salmon farming and account for most of the pollution around fish farms. The organic waste, which comes primarily from fish faeces and feed waste, can build up on the seabed if the rate of decomposition is sufficiently low, thereby damaging the local fauna. Another problem is that the waste leads to higher concentrations of nutrients in the sea, which increase the risk of eutrophication. However, eutrophication depends on the nutrients being discharged and on the resilience of the local environment. A strong current increases the availability of oxygen, which is needed for decomposition of the organic matter, and also contributes to its wider dispersion. Hence, the organic load directly under the cages is reduced, thereby alleviating the challenge to the environmental resilience capacity. Since seabed topography also influences the resilience of the environment, the siting of cages is important.

Organic waste sedimentation not only poses a problem for the local fauna, but also for salmon farmers due to negative feedback effects on productivity. The biological decomposition process for the waste reduces the availability of oxygen in the surrounding area, thus lowering the resistance

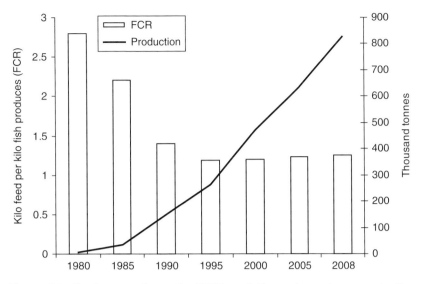

Figure 5.5 Feed conversion ratio (FCR) and Norwegian salmon production, 1980–2008. (*Sources*: Austreng (1994), Norwegian Directorate of Fisheries)

of farmed fish to diseases. Moreover, depletion of the oxygen level in the decomposition process can produce toxic gases which, if released, are harmful to farmed fish. Thus production risk increases with higher feed use because of the negative environmental feedback. Therefore, risk-averse salmon farmers would minimise feed use and/or take other measures to reduce negative feedback effects on productivity. As feed costs may account for over 50% of the total production costs in salmon farming, there is also a cost argument for reducing feed use.

Salmon farmers have responded to these problems. First, feed and feeding technology have improved considerably over the last two decades. Figure 5.5 shows that the feed conversion ratio (FCR) declined from almost 3 kg of feed required to produce 1 kg of salmon in 1980 to just over 1 kg in 1995, after which time it has levelled off. Most of this reduction is due to a greater use of lipids in the feed: a 1% increase in the inclusion rate of lipids leads to a 1% reduction in organic waste. However, new feeding systems have also contributed to reducing the FCR by lowering feed waste.

Second, most salmon farms are now located in areas with relatively strong currents, deep water and suitable seabed topography. This significantly reduces the accumulation of waste sediments and negative feedback effects on productivity. In areas with unsustainable locations, salmon farms have disappeared. Thus, the combination of new sea cage technology, which allows sites to be moved to more exposed locations, rotation between different sites, and improved feed and feeding technology has significantly increased the elasticity of substitution between traditional factors of production and effluent discharges. This undertaking has probably been

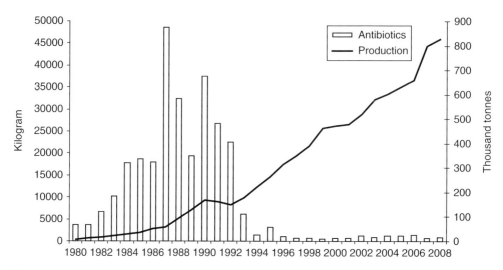

Figure 5.6   Use of antibiotics and Norwegian salmon production, 1980–2008. (*Source*: Norwegian Directorate of Fisheries)

induced by a combination of environmental feedback effects on productivity and a general effort to reduce costs. Consequently, salmon farmers have internalised many of the problems related to organic waste so that the environmental quality of the areas surrounding the salmon farms has improved since the late 1980s.

In this context, it is interesting to note that the improvements in feed and feeding regimes have, to a large degree, been made by the feed industry, not by the salmon farmers. Hence, this is an example of how the suppliers to salmon farmers contribute not only to increased productivity but also to improved environmental production practices.

5.2.2   *Antibiotics and chemicals*

The use of antibiotics in the treatment of diseases is another controversial issue concerning the environmental practices of salmon aquaculture. Antibiotic use can lead to antibiotic resistance in fish and other living organisms. In particular, the extensive use of antibiotics in the late 1980s provoked much criticism from consumers. Since then, the use of antibiotics has been virtually eliminated due to the use of vaccines.

Figure 5.6 shows that the use of antibiotics forms an inverted U-shaped pattern. First, salmon farmers responded to the disease problem in the 1980s by increasing the use of antibiotics. The first large disease outbreaks were bacterial cold-water vibriosis in 1986 and furunkulosis in 1990–1992. Two factors were important in reversing the trend towards increasing use of antibiotics. First, the relocation of salmon farms to more suitable locations

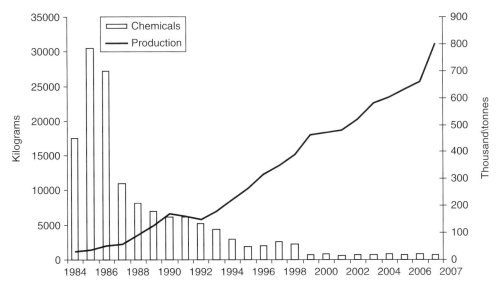

Figure 5.7   Use of chemicals and and Norwegian salmon production, 1984–2007. (*Source*: Norwegian Directorate of Fisheries)

generally improved fish health. Second, the introduction of an oil-based vaccine in 1992, which was effective against bacterial diseases, made antibiotics more or less redundant. Thus, since peaking in 1987, the use of antibiotics has been on a downward trend, despite a temporary increase in usage following the furunkulosis outbreaks in 1990. This contrasts with the upward-sloping trend for production. After the first vaccinations took effect in 1993, antibiotics were hardly used.

The development of the oil-based vaccine can be seen as the result of the salmon industry becoming an attractive market for industry-specific pharmacy services and products. Industry growth therefore made it profitable for the pharmaceutical industry to invest in the development of such vaccines, which would otherwise not have been available until much later. Thus, industry growth has helped to reduce the use of antibiotics, not only in relative terms but also in absolute terms.

The same overall trends are found in the use of chemicals. Figure 5.7 shows that since the mid 1980s, the use of chemicals has demonstrated a downward trend. Because the time series only dates back to 1984, we only observe the downward-sloping trend in the use of chemicals. However, we can infer an inverted U-shaped pattern for chemicals, given that their use must have been close to zero in the 1970s before intensive salmon aquaculture began. Chemicals are mainly used for cleansing cages and for treating salmon lice. Wrasses have been introduced as a more environmentally friendly method of treating sea lice because they feed on the lice that live on salmon. On its own, this measure is not sufficient to eliminate sea lice.

Salmon farmers must still rely on chemicals to treat infected fish, but they use considerably less now than they did in the mid 1980s. Yet, as in the case of antibiotics, we observe a decline in the use of chemicals as the salmon industry expands.

5.2.3   Salmon escapees and sea lice

The issue of salmon escapees is controversial because of its potential negative impact on wild salmon stocks. The short-term effects of escaped farmed salmon include competition and breeding with wild salmon, the spreading of diseases and parasites to wild salmon, and hybridisation with trout. Since a number of theories have tried to explain why wild salmon stocks have been reduced, the actual effects of farmed salmon on wild salmon are still open to question.[15] Nevertheless, escaped farmed salmon in large numbers probably have a negative impact on wild salmon stocks.

The main reasons for accidental release of farmed salmon are winter storms, propeller damage, and wear and tear on equipment. In recent years, better management of these problems has ensured that the number of salmon escapees is relatively stable, which contrasts with the increased number of salmon produced each year. In Figure 5.8, the number of escapees is shown for the period 1988–2008, Note that the figures for 1988–1992 are mean figures for this period, together with total production. As can be seen, before 2006, for a long time the highest number of salmon escapees in a single year was in 1994 with 671 000 individuals. For the period under consideration, there appears to be two 'regimes' for the number of escapees. In about two-thirds of the years, there are 200 000–400 000 escapees, while in the remaining years there are 600 000–900 000. A new record was set in 2006 with 920 000 escapees. However, numbers for 2007 and 2008 are among the lowest number of escapees reported, so this does not seem to be an increasing trend. Rather, it seems like the risk has increased with the higher production, so that one may expect that the number of escapees in difficult years increases, but that it remains stable in good years. This is as expected given that the size of both farms and pens increase and, accordingly, each accident leading to escapees is likely to involve a larger number of fish.

---

[15]   There is no doubt that wild salmon stocks are in trouble many places. In some cases, like wild salmon in Spain and France but also some river systems in Norway and Scotland, it is clear that the problems were unrelated to salmon aquaculture, as the stocks had largely disappeared before salmon aquaculture commenced (and salmon aquaculture is not conducted in Spain and to a very limited extent in France). For these stocks, development in the rivers (road and power station construction, acid rain) as well as oceanographic changes are mentioned as the main factors. These factors are certainly also present farther north, but there are also significant numbers of escaped farmed salmon in many rivers. Hence, it is certainly possible that escaped farmed salmon can have an impact on wild stocks in salmon-producing areas.

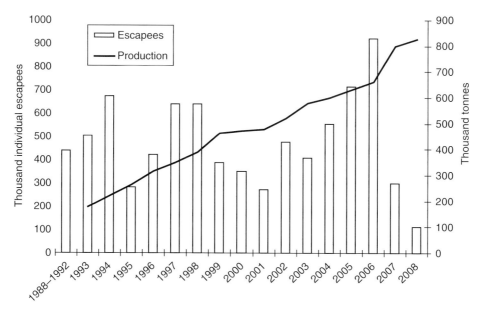

Figure 5.8   Number of escaped salmon and Norwegian salmon production, 1988–2008. (*Source*: Norwegian Directorate of Fisheries)

Nevertheless, these figures should be treated with some caution, because they are probably lower than the actual number of escapees. Since escapes of salmon can generate negative publicity and may lead to lawsuits, salmon farmers have incentives to under-report. Farmers may also be unaware of escapes because damage to cages is detected late, or they may not know exactly how many fish were in the cages. However, under-reporting is unlikely to affect the main trends.

Systematic fishing by net in set locations along the Norwegian coast and rivers provides another indicator of the escaped farmed salmon situation. These figures show that the share of farmed salmon in catches has decreased from 1989 to 2005. Importantly, the share of farmed salmon in broodstock rivers for wild salmon has decreased from 39% in 1989 to 13% in 2003.

Infection by sea lice is possibly an important factor in the reduction of wild salmon stocks in salmon-producing regions. Registrations show that the heaviest infections of wild salmon are limited to areas with a high concentration of salmon farms. A plausible explanation is that the number of hosts is larger in areas with a high concentration of salmon farms, thus leading to a higher concentration of sea lice in that area. Moreover, escaped salmon are believed to be one of the major causes of spreading salmon lice to wild salmon.

Sea lice infections and salmon escapes are probably the major remaining environmental problems in salmon farming today. Salmon farmers clearly have an incentive to limit the number of sea lice because of negative feedback effects on productivity and for marketing purposes. However, it is not clear that salmon farmers have an incentive to reduce the number of sea lice to a level that is significantly below the level required by the market. This means that sea lice concentrations in salmon farming areas might be relatively high even if the number of sea lice living on the farmed salmon is at an acceptable level. Thus, it is uncertain whether there has been an inverted U-shaped pattern for the level of sea lice in salmon farming areas. Research on vaccines against sea lice continues, but there has been no major breakthrough to date.

5.2.4 Concluding remarks

Large-scale aquaculture has the potential to severely impact its local and regional environment. However, the control of the production process that makes large-scale intensive production possible can also be used with regard to environmental matters. As an issue has to arise before it can be addressed, it can take some time before one is able to respond. In Norwegian salmon aquaculture, the environmental issues have to a large extent been addressed. This is partly because the negative environmental influences also have a negative impact on production costs and partly because of regulations. We do not have data to make any direct assessments about other countries, but as there are a number of multinational companies in the production as well as the feed sector, there is strong reason to believe that these practices are transferred. Industry sources indicate that this is the case. Furthermore, there is also some form of environmental regulation in all the salmon-producing countries, limiting what the industry can get away with.

   The state of environmental practices for salmon should hold an important lesson also for other species, even though all issues have not been fully resolved. With a sufficient degree of control of the production process, all environmental issues can be resolved, and salmon aquaculture is certainly sustainable. It will not necessarily be the case that the industry will have incentives to internalise all environmental issues, and as such there is an important role for regulatory authorities. However, to the extent that aquaculture is carried out in an unsustainable manner, it is not because it is intrinsically unsustainable. It is rather because authorities are not able or willing to implement and enforce regulations that make the producer use the best practices available.

Bibliography

There are a number of papers discussing potential environmental issues of aquaculture. Naylor *et al.* (2000) provide a survey from a critical perspective. Asche and Tveterås (2004) and Kristofersson and Anderson (2006) investigate whether the fish meal trap hypothesis seems to hold. Hannesson (2003) presents a model of the interaction between a fish stock used as feed in aquaculture and the demand for feed from aquaculture. For an overview of different fisheries management regimes, see Munro and Scott (1985) and Bjørndal and Munro (1998). Bjørndal (1988) analyses the management of fisheries for pelagic species. Arrow *et al.* (1995) provide a more general discussion of economic growth and sustainability.

There are a number of studies of local environmental effects of aquaculture in general, and salmon aquaculture in particular. These include the discussions between Folke *et al.* (1994) and Black *et al.* (1996), as well as Wallace (1993), Grimnes *et al.* (1998), Asche *et al.* (1999a), and Tully and Nolan (2002), Tveterås, S. (2002) and Holmer *et al.* (2008). Olaussen (2007) looked at recreational salmon fishing.

6 Markets for Salmon

Seafood is consumed all over the world. However, consumption varies substantially depending on geographical location, culinary tradition and ability to pay. Although prices for salmon have been decreasing, it is still a relatively expensive product. The ability to pay therefore makes the EU, Japan and the USA the most important markets, and this is where the most significant quantities are consumed. However, good logistics and a positive reputation make salmon a species that is consumed all over the world (Norway and Chile combined export to almost 150 countries). Moreover, there is also significant development in several emerging markets such as Brazil, Russia, eastern Europe and Southeast Asia, where consumption has increased significantly since 2000.

As discussed in Chapter 1, during the last 50 years the world's seafood consumption has increased, and since the 1970s aquaculture production has contributed substantially to this increase. Figure 6.1 shows per-capita consumption of fish in selected countries for the period 1961–2005. The figure gives two main insights. The per-capita consumption increased throughout the period in the world as a whole as well as in the most important industrialised countries. Moreover, per-capita consumption in Japan was much higher, and increased from about 50 kg in the early 1960s to about 70 kg in 1973. It remained at that level with substantial year-to-year variation until the turn of the century before it started to decline; in 2005 it was estimated at 61.2 kg. In the other large industrialised countries, the consumption varies substantially despite the common trend. In 2005, per-capita consumption for the USA was 24.1 kg, 14.8 kg for Germany and 35.3 kg for France. Despite the fact that the global per-capita consumption is also increasing, it still remains low compared with most industrialised countries at 16.4 kg per capita in 2005.

Japan has the highest seafood consumption per capita, and with its relatively large population it is the single largest fish importer in the world with an import value of US$14.3 billion in 2006, followed by the USA at

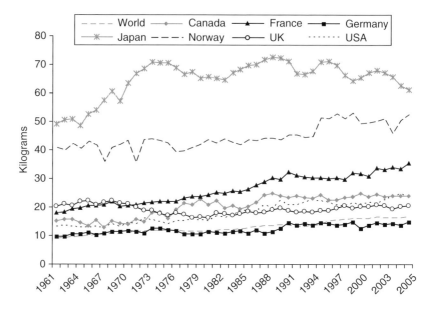

Figure 6.1 Per-capita seafood consumption (kg) in selected countries, 1961–2005. (*Source*: FAO)

US$13.4 billion.[16] However, in contrast to most other markets, consumption and imports to Japan have been stagnant during the last decade. With a strong tradition of salmon consumption, Japan is also the country with the largest per-capita salmon consumption in the world, although the EU became the largest market in the 1990s. Spain is the largest market for fish and fishery products in the EU, followed by France. However, France is the largest European importer of seafood, and the most important European market for salmon.

6.1 The European Union markets

In the early 1980s, when the production of farmed salmon started to grow, there were two main markets for salmon in the world, Japan and the USA, while the European market was negligible. This has changed substantially as the European salmon market has grown tremendously, primarily consuming fresh farmed salmon. In the mid 1990s, the EU became the largest single salmon market. In 2008, total import value was €2.1 billion, comprising 571 000 tonnes in product weight and over 687 000 tonnes in round fish

---

[16] However, the EU as a group is even larger, with an import value of about US$37.8 billion in 2006.

equivalents.[17] In addition, there is the Scottish and Irish production consumed within these countries that will add about 70 000 tonnes. With a similar unit value as for the imports this would add about €250 million to the market value. As such, the EU market at the wholesale/import level is about 760 000 tonnes round fish equivalents with a value of about €2.35 billion. Norway is the largest supplier, with an export value of €1.6 billion, comprising 534 000 tonnes in round fish equivalents in 2008. At the import stage, whole fresh salmon is the dominant product form, with an import share of about 65%, followed by frozen fillets with an import share of about 20%.

Despite the tremendous growth in the consumption of farmed Atlantic salmon, the EU also continues to import frozen wild Pacific salmon. In 2008 the imports were at 18 700 tonnes. During the last decade, imports of Pacific salmon have been at a similar level to that in the early 1980s. Hence, Pacific salmon has not to a large extent lost markets in Europe despite the formidable growth in the imports of farmed salmon. The leading importers are France and Germany.

The EU market is heterogeneous as the different countries have substantially different seafood consumption traditions and different access to seafood. Spain is the largest seafood market, while France is the largest importer and Portugal has the highest per-capita consumption. In some of the new inland member countries, like the Czech Republic, there is not a strong seafood tradition, while for others EU membership gives new opportunities. For instance, Poland has in a few years built up a substantial processing industry based on salmon imported from Norway. We will here look closer at the largest salmon-consuming countries in Europe.

6.1.1 France

Overall, France is the second largest market for fish and fishery products in the EU, and the largest importer. The average consumption of seafood per capita was 35.3 kg in 2001, well above the average in Europe as a whole, and ranking France fourth among the EU countries. France is also the largest European market for salmon. Counting all product forms, salmon constitutes about 10% of French fish consumption, equivalent to an annual per-capita consumption of roughly 2 kg salmon.

A substantial share of seafood consumed in France is imported. Total salmon imports have been increasing since the 1980s to about 158 000 tonnes in round fish equivalents in 2008. In particular, the imports of whole fresh Atlantic salmon have increased from about 6000 tonnes to 89 800 tonnes in 2008. During the last few years imports of fresh salmon seem to have stabilised at a little over 80 000 tonnes. In contrast, imports of frozen Pacific salmon

---

[17] A round fish equivalent (or wild fish equivalent) is the round (unharvested) weight of a harvested fish. Conversion factors vary from species to species and between sources, but for salmon 0.9 for whole fresh and frozen and 0.65 for fillets is common.

Table 6.1   French imports of salmon (thousand tonnes product weight).

	Fresh whole Atlantic	Frozen whole Atlantic	Frozen whole Pacific	Fresh fillets	Frozen fillets
1985	6.3	16.3	–	0.9	–
1990	40.5	14.5	13.2	0.3	1.2
1995	64.9	4.3	12.5	1.3	4.6
1998	73.1	1.5	6.9	4.4	6.8
1999	88.5	1.4	9.0	5.6	8.3
2000	74.6	1.1	8.1	3.6	10.6
2001	82.1	1.5	9.2	1.6	11.1
2002	81.2	1.7	7.6	1.5	10.9
2003	91.1	2.0	7.1	2.0	12.1
2004	80.3	2.3	7.7	5.5	13.5
2005	87.8	4.1	6.1	4.9	15.9
2006	84.9	4.8	6.6	5.1	18.2
2007	85.8	4.6	6.4	4.9	18.7
2008	89.7	3.1	5.6	6.1	19.5

Sources: FAO, OFIMER.

have dropped somewhat, from about 15 000 tonnes in the 1980s to about 6000 tonnes in 2008. There has also been a large increase in imports of salmon fillets, especially frozen during recent years. This is a trend that is likely to continue. French salmon imports by product form are shown in Table 6.1.

France imports salmon predominantly from European producers. Norway is the main supplier of Atlantic salmon with about 60% of the imported volumes, with the UK second largest with about 20%. In recent years smaller quantities of frozen Atlantic salmon, and particularly fillets, have been imported from Chile, making Chile the third largest supplier. In the French frozen food service market Chilean producers have succeeded in establishing a growing presence, and Chile represents the main supplier of frozen salmon fillets.

Imported salmon are divided roughly equally among retail sales, restaurant/catering and the processing industry, which consists mainly of smokehouses, but increasingly also of other types of secondary processing such as ready-made meals.

Salmon is the principal fresh fish species consumed at home, with about 15% of the home consumption market for fresh fish. Fresh and smoked salmon together attain around 30% of the market value for fresh and smoked fish. The retail chains dominate sales of fresh salmon to households, with over 80% of the purchase value. Fresh salmon consumption is shared equally between home consumption and the restaurant/catering industry. Of the latter, restaurants account for about 75% of consumption and the catering industry for the remaining 25%.

The French smoking industry utilises around 35 000 tonnes of imported salmon per year; 95% of the raw material is farmed salmon. Industry output increased steadily in the 1990s and is currently over 20 000 tonnes per year, of which the bulk is sold through the retail sector. Most of the production is sold in the domestic market and only about 10% is exported, mainly to Italy and Belgium. The demand for smoked salmon is seasonal, although less so than in the past, with Christmas/New Year and Easter being the peak seasons.

The smoking industry is a traditional industry that experienced a tremendous growth in the late 1980s as the use of farmed Atlantic salmon and lower prices provided a very popular product. After very good years in the 1980s, the industry suffered several setbacks in the early 1990s and experienced major structural changes during the last half of the 1990s. Quality had deteriorated as injecting brine and smoke flavour had become commonplace, and subsequently prices plummeted. Since then, extensive consolidation has taken place, and less than 10 smoking companies now dominate retail sales. Several of these companies are foreign owned, and Marine Harvest is one of the largest producers of smoked salmon in France. The focus on improving product quality and regaining consumer confidence has been largely successful together with a strong focus on branding of products, and the industry has experienced considerable sales growth since the turn of the century. However, as the smokers are caught in the middle between the producers and the marketers, they often experience substantial pressure on their margins.

The great success of salmon in the French market can largely be attributed to the stable supply throughout the year and a wide selection of high-quality products sold at reasonable prices, due to technological advances and economies of scale in the production and distribution of farmed salmon. This has also led to a substantial increase in the product forms being consumed, as lower prices have led to significant product innovation. As the market matures, consumers become more focused on product diversity, geographic origin, production process and quality standards. In general, Scottish and Irish farmed salmon have been perceived to be of higher quality than the Norwegian product (this is also manifest in the form of higher prices). However, consumers have become increasingly aware of the salmon farming process. As a result of negative publicity, farmed salmon lost some of its natural image, giving organically farmed salmon and wild salmon, as well as other wild species, a potential advantage. In the future, it therefore seems reasonable to expect an increased importance for product labelling and greater focus on geographic origin and mode of production. However, the main trend seems to be an increased number of product forms, where most are value-added forms that make the product easier to consume. As the introduction of new product forms

Table 6.2   German imports of salmon (thousand tonnes product weight).

	Fresh Atlantic	Frozen Atlantic	Fresh fillets	Frozen fillets
1985	3.5	–	–	–
1990	13.4	1.9	–	–
1995	54.2	2.3	2.2	5.0
2000	57.7	4.1	4.1	8.0
2001	53.2	5.0	5.0	10.8
2002	50.7	3.5	4.2	13.0
2003	47.2	4.1	5.2	17.8
2004	50.4	4.9	6.2	22.3
2005	56.1	5.5	6.8	26.6
2006	47.3	6.2	6.6	33.5
2007	47.0	5.6	6.2	34.5
2008	38.0	5.3	6.2	33.7

Sources: FAO, Eurostat.

also seems to increase market size, this creates a more differentiated market and fuels an increase in the total demand for salmon.

6.1.2   Germany

Germany is the EU country with the largest population, and although the per-capita seafood consumption is not among the highest, this makes Germany the second largest market for salmon in Europe. Table 6.2 shows the supply of salmon to the German market for 1985–2008. While fresh Atlantic salmon in most years has been the most important product form, it is interesting to note that the quantities have been relatively stable since the mid 1990s. The growth has been for other product forms, and particularly fillets. The most striking feature is the increase in the supply of frozen salmon fillets. When converted to round weight equivalents, frozen fillets were more important than whole fresh salmon in 2008.[18] Germany continues to import a small amount of frozen Pacific salmon, totalling 7100 of the 7200 tonnes of whole frozen salmon that was imported in 2008, mainly from the USA.

While Norway is the main supplier, a substantial share of the salmon originating in Norway is shipped via a third country, where additional processing takes place. For fresh and frozen salmon fillets, Denmark traditionally has had a leading role. However, in recent years, Poland has become an important source for processed salmon products. In some years

---

[18]   33 700 tonnes of frozen fillets is 51 900 tonnes in round fish equivalents, while 38 000 tonnes whole fresh salmon is 42 200 tonnes round fish equivalents.

Chile is an important source for frozen fillets, particularly when prices in Europe are relatively high.

Germany is a significant importer of smoked salmon, with imports of about 22 000 tonnes in product weight. Denmark has traditionally been the dominant supplier, but in recent years Poland has overtaken this position. Germany also has several domestic smokehouses relying on fresh imported raw material from Norway and the UK, the largest being the Bavarian-based Laschinger Group.

Salmon is consumed both at home and in restaurants. Organic salmon is a growing niche, although volumes are still small. Germany is a very price conscious market and the relatively low price of salmon is seen as a key factor in increasing sales further. Frozen salmon products and especially portions are widely distributed through national discount chains such as Aldi. Supermarkets and hypermarkets have increased their share of salmon sales and with the recent liberalisation of shop opening hours, salmon sales through the retail chains are expected to grow. While the imports of salmon have a very different structure from those in France, this is to a much lesser degree reflected in the consumption patterns. If anything, French consumers tend to buy more highly processed product than German consumers. However, a much larger share of the processing takes place in France.

6.1.3   United Kingdom

The UK produced almost 140 000 tonnes of salmon in 2008 (see Chapter 3). UK exports have largely followed production and increased steadily until 2003. However, while exports have been important, the domestic market has always been the most significant market for Scottish producers. Moreover, this has made the UK market one of the largest salmon markets in Europe, with consumption of about 120 000 tonnes as measured in round fish equivalents (and disregarding tinned salmon). This makes salmon the largest species in the UK home consumption market for fresh and chilled fish. Also in the UK, the smoking industry is important. About 30 000 tonnes of fresh salmon is used in smoked salmon production, with a 40 : 30 : 30 split between in-house consumption, food service segment and exports.

UK demand for salmon remains strong even though Scottish production peaked in 2003. This has led to a substantial increase in imports, as can be seen in Figure 6.2 where import quantity is shown together with the export quantity measured in round weight equivalents. During the last few years, imports reached record levels at over 50 000 tonnes, and net exports were only 14 500 tonnes in 2008. As one can see, also in 1999 and 2000 imports were significant, reaching 30 000 tonnes. This was a period with generally high prices, where Scottish salmon seemed to get a higher price premium relative to Norwegian salmon than usual. Scottish exports to overseas markets, and particularly the USA, increased substantially in this period. Imports were then reduced until 2003 when Scottish production reached a

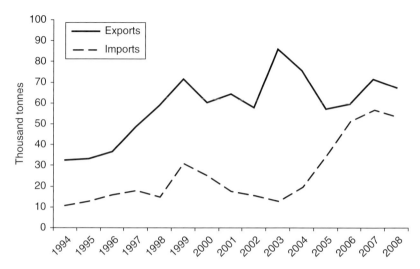

Figure 6.2   UK salmon exports and imports, 1994–2008. (*Source*: Eurostat)

peak, and after that imports have increased rapidly. The main source for imports is Norway with a share of over 80%. The fish are either imported directly or via Sweden, Poland or Denmark. The Faroe Islands and Chile are the second and third most important exporters, but only a few hundred tonnes in product weight were imported from Ireland.

The UK continues to be the most important export market for tinned salmon from the USA. With no domestic production, the UK is almost wholly reliant on the USA and Canada for its supplies of tinned Pacific salmon. Consumption of tinned salmon is mainly by older generation consumers. Despite some year-by-year oscillations, demand is holding up quite well.

6.1.4   Spain

Spain is the largest seafood market in Europe, and one of the markets where salmon is relatively less important. Nevertheless, the size of the market makes Spain an important market as imports of salmon have picked up. The supply of salmon to Spain increased considerably from the beginning of the 1990s (Table 6.3) and in 2008 44 500 tonnes were imported when measured in round fish equivalents at a value of €134 million. Salmon is mainly imported as a fresh product, and the import share of whole fresh salmon was 71% in 2008. While salmon is mostly consumed as a fresh product, a significant share also goes to domestic processors like the smokehouses. Whole frozen salmon is the second most important product form with an import share of 20%. This is evenly split between farmed Atlantic and wild Pacific salmon. Hence, imports of other product forms are

Table 6.3   Spanish imports of salmon (thousand tonnes product weight).

	Fresh Atlantic	Frozen Atlantic	Fresh fillets	Frozen fillets
1985	1.5			
1990	14.5	0.3		
1995	16.5	2.6	0.1	0.4
2000	21.7	5.0	0.5	0.3
2001	32.2	5.5	0.3	0.5
2002	32.0	4.7	0.3	1.0
2003	33.5	5.6	0.2	1.0
2004	24.9	4.8	0.3	1.5
2005	28.9	4.3	0.5	1.5
2006	26.6	7.5	0.4	1.9
2007	28.1	1.5	1.2	3.6
2008	26.9	0.9	2.1	2.3

Sources: FAO, Eurostat.

relatively limited. This is to some extent also reflected in consumption patterns, as the product range is not as wide as in France. Norway is the main supplier, with an import share of almost 90%.

As noted, there has been a substantial increase in imports in most years since the early 1990s. There are several reasons why the market for salmon has grown slower in Spain than elsewhere in Europe. These include the longer distance from the salmon-producing countries in northern Europe, the much higher share of seafood that is still sold in traditional outlets and the larger supply of seafood from domestic fishermen.

6.1.5 *Italy*

The supply of salmon to Italy is given in Table 6.4; in 2008, salmon consumption was 41 234 tonnes measured in round fish equivalents. Norway is the largest supplier, with a market share for fresh Atlantic salmon of almost 80%; as the remainder mostly comes from Denmark and recently Poland, most of that is also from Norway. Italy is the most important market for smoked salmon for the Danish smoking industry, with an import quantity of about 4000 tonnes in 2008.

Italy is a large importer of fish and fishery products and ranks today as the fifth largest importer in the world. Salmon consumption and imports have grown during the last decade in line with this increase, but the main increase in farmed finfish consumption has been Mediterranean sea bass and sea bream, primarily from Greece but also from domestic sources.

As in most other European countries, the bulk of salmon sales are through supermarkets and hypermarkets. Although fish consumption in Italy is

Table 6.4   Italian imports of salmon (thousand tonnes product weight).

	Fresh Atlantic	Frozen Atlantic	Fresh fillets	Frozen fillets
1995	8.7	2.1	0.1	0.4
2000	10.7	2.4	1.5	1.3
2001	12.0	2.4	1.8	2.0
2002	12.7	1.9	1.9	2.2
2003	13.3	2.1	2.1	1.9
2004	13.1	2.8	2.3	2.1
2005	13.2	2.5	3.0	2.0
2006	13.6	3.2	2.8	2.3
2007	14.9	1.4	1.2	3.6
2008	16.0	0.9	2.1	2.4

Source: Eurostat.

higher on the coast, the islands and in the south, salmon consumption is concentrated in northern Italy and is only slowly penetrating the rest of the peninsula, although significant quantities are now consumed in the larger cities in the south.

6.1.6   Denmark

Denmark does not produce farmed salmon itself but is an important trader of fresh salmon from Norway and the Faroe Islands. This makes Denmark not only a significant importer but also the second largest exporter of salmon in Europe after Norway. Danish imports are shown in Table 6.5. Over time, Denmark and France have in most years alternated as the number one location with respect to being Norway's largest export market for salmon (Japan also held this honour for a few years). In 2008 total imports in round weight equivalents were 185 800 tonnes at a value of €383 million. Danish imports peaked in 2003 at 218 100 tonnes round fish equivalents. Since then increased competition from processors in other EU countries, particularly Poland, has won significant market shares from the Danish industry.

Denmark's salmon imports are fairly evenly divided into fish that are re-exported with little or no processing and fish that are used for processing. In 2008 the export value for unprocessed fresh salmon was €229 million, while the export value of all salmon was €397 million. Its processing industry and smokehouses import raw material that is re-exported as processed products all over the world, but especially in Europe. This is because non-EU processors are less competitive since tariffs on processed products are higher than for unprocessed products. Denmark is thus a

Table 6.5   Danish imports of salmon (thousand tonnes product weight).

	Fresh Atlantic	Frozen Atlantic	Fresh fillets	Frozen fillets
1991	12.3	16.8	0.8	0.9
1995	74.1	5.7	2.4	3.3
2000	139.2	3.9	6.2	8.4
2001	140.5	4.4	6.8	12.0
2002	153.0	3.9	6.8	9.4
2003	161.7	3.8	9.0	10.6
2004	143.0	2.9	10.9	10.8
2005	122.1	4.0	10.2	19.2
2006	115.5	4.2	6.6	15.2
2007	129.5	4.7	10.6	20.2
2008	126.9	2.5	8.5	18.6

Source: Eurostat.

major producer and exporter of smoked salmon and of portions and fillets. Smoked salmon is the most important product form, with an export value of €50 million in 2008. In the market for smoked salmon, Denmark has been a very competitive producer, but rising labour costs have moved some operators to set up production in countries such as Poland and Estonia after these countries joined the EU.

The local market is limited because of the small population, but the per-capita consumption of salmon is relatively high.

6.1.7   Poland

The Polish market is expanding rapidly as shown in Table 6.6. Imports in round fish equivalents have increased from less than 5000 tonnes in 2000 to over 80 000 tonnes in 2008. Norway is the largest exporter, with a market share of more than 90%. In 2008 Poland became the second largest market for Norwegian salmon, taking over this spot from Denmark. Poland differs from most other markets in that almost 90% of the imports were whole fresh salmon. The main reason for this is that Poland, in addition to being a substantial market in its own right, has also developed a substantial processing industry. This industry is highly competitive, and a strong competitor to the export-oriented Danish processing industry as well as the domestic industries in other EU countries. Germany is the main market for the processed re-exports, with smoked salmon as the main product. In 2008 the total export value was €54 million.

The consumption of salmon in Poland, as in most eastern European countries, is highest in the large cities. It is also in and around the large

Table 6.6   Polish imports of salmon (thousand tonnes product weight).

	Fresh Atlantic	Frozen Atlantic	Fresh fillets	Frozen fillets
2000	5.6	0.4	0.0	0.2
2001	8.4	0.3	0.0	0.4
2002	12.3	0.4	0.1	1.4
2003	20.1	0.4	1.0	1.3
2004	24.5	0.4	6.7	1.4
2005	38.1	0.4	5.6	4.0
2006	44.7	0.4	6.5	3.3
2007	51.3	0.5	7.7	2.3
2008	65.8	0.3	0.2	2.4

Source: Norwegian Seafood Exports Council.

cities that the big supermarket and hypermarket chains have their outlets. Since the middle of the 1990s there has been a substantial growth in this segment, and in 2006 Poland was one of the countries in Europe with the highest share of food retail sales through the supermarket chains. In the period 1995–2001 the number of hypermarkets increased from 10 to 136, and in 2000–2001 alone grew by 21%. By the end of 2005, 200 hypermarkets and 1400 supermarkets were operating.[19] Many of the largest chains are foreign owned, for example Tesco of the UK and Auchan of France.

6.2   The Japanese salmon market

Total Japanese salmon consumption (round weight equivalent) has been at about 500 000 tonnes every year since the early 1990s. The composition is split approximately fifty-fifty between imports of mostly high-value salmon (Atlantic, coho, salmon trout and sockeye) and domestically caught low-value salmon (chum and pink). A very wide variety of salmon species and products are consumed in Japan (for an excellent and thorough appraisal of the Japanese seafood market see Nakomoto 2000). Salmon consumption varies by geographical area and time of year. Since salmon has traditionally been caught in the northern parts of Japan, salmon consumption is significantly higher there than in the south. Moreover, as all salmon species are landed or imported to Japan, it is the most diversified salmon market in the world.

Historically, Japanese salmon consumption has been highly seasonal. Japanese fish consumption has always fluctuated with the local availability

---

[19]   *Source*: CAL Company Assistance and www.retailpoland.com

of different species, as well as their changes in quality throughout the year. For example, autumn chum salmon consumption is concentrated during the harvest period between September and December. However, seasonal consumption patterns have weakened somewhat over time as freezing technology allowed salmon to be consumed at times other than during the run, imports have expanded the times at which wild species are available, and farmed salmon and trout have become available year round.

Another important factor contributing to seasonal fluctuations in salmon consumption is the celebration of different holidays and major social events. Sending gifts is often an important part of various celebrations, and salmon products are highly rated gifts. Except for smoked salmon, farmed salmon do not generally make popular gift items because of the perishability.

The preparation of salmon varies by species. Grilling is the most common preparation of salmon in Japan. Fillets of smaller fish are grilled whole, and fillets of larger fish are cut into rather small slices, called *kirimi*, which may be natural, salted or marinated. Grilled salmon is served in a number of different ways. Unsalted salmon may also be broiled with a sauce, for instance *teriyaki* sauce, or pan-fried with butter in a 'western' preparation. Smoked salmon remains a high-end product, typically consumed in hotels and at banquets.

Traditionally, salmon was not consumed raw in Japan, because of the presence of parasites in wild salmon. However, since the introduction of farmed salmon, consumption of raw salmon has increased. Salmon *sashimi* (sliced raw fish) is widely available in supermarkets, restaurants and sushi bars. *Aramaki* style salmon is head-on gutted fish heavily layered in granular salt. *Aramaki* style salmon was formerly a major product form, particularly for autumn chum salmon, but has declined in importance as incomes have increased and food preparation time has decreased.

Usage and preparation of salmon differs by species depending on the texture of the meat, the oil content and the colour. While wild salmon was traditionally preferred, farmed salmon has gained increasing acceptance as farmed supply has expanded. Wild sockeye has been preferred for its intense red colour and is still highly appreciated in the Kansai region of southern Japan. Sockeye is most commonly sold as salted *kirimi*. However, due to poor sockeye catches in recent years, many Japanese have become used to farmed coho or salmon trout, which both offer the same reddish flesh colour, as substitutes for wild-caught sockeye. Farmed coho has benefited from the long presence of wild coho in the Japanese market. Similar to wild sockeye, coho is also most commonly sold as salted *kirimi*.

Salmon trout is the most versatile species in the salmon market. Appreciated for its red flesh and high oil content, salmon trout is sold salted as *kirimi*, raw as *sashimi* or *sushi*, defrosted for pan-frying, marinated, or processed as smoked trout. For its many uses, salmon trout has gained a strong position on the Japanese market. Airfreighted chilled Atlantic

Table 6.7 Japanese salmon imports, 1993–2008 (tonnes product weight).

Species	1993	1995	2000	2001	2002	2003	2004	2005	2006	2007	2008
Fresh product											
Total	13 688	18 376	26 429	31 453	28 716	27 418	27 353	26 302	23 016	22 506	20 689
Atlantic	10 763	16 065	24 992	29 621	27 154	25 233	25 28	24 817	21 916	21 577	19 991
Other	2 925	2 311	1 437	1 832	1 562	2 185	2 073	1 485	1 100	929	698
Frozen product											
Total	214 851	181 557	196 693	242 772	256 151	203 363	233 581	217 519	203 448	199 074	211 143
Atlantic	4 714	7 385	6 625	7 836	6 309	3 842	4 732	5 563	2 856	2 38	4 362
Trout	15 787	30 476	58 639	82 99	98 177	82 709	86 669	70 989	72 807	73 071	79 873
Coho	24 984	41 265	64 759	88 577	81 457	57 863	77 102	72 405	73 382	72 681	81 807
Sockeye	132 639	90 556	52 651	49 633	55 046	47 546	51 525	55 679	44 755	45 731	40 622
Other Pacific	3 648	11 875	14 019	13 736	15 162	11 403	13 553	12 883	9 648	5 211	4 479
Fresh and frozen total	228 539	224 057	249 551	305 678	297 413	241 697	265 07	249 862	223 077	217 865	226 378
Trout fillets	11 702	15 769	12 920	22 106	16 170	16 502	23 217	20 261	26 403	26 221	26 143
Total including fillets	240 241	239 826	262 471	327 784	313 583	258 199	288 287	270 123	249 48	244 086	252 521

Notes: For fresh salmon, 'other' species are mostly wild but include small volumes of farmed trout. 'Other Pacific' salmon includes chinook, chum and pink salmon. 'Other Pacific' salmon from Canada is mostly wild but may include small volumes of farmed chinook. Trout fillets from Chile, reported in the penultimate row of the table, are the only significant component of salmon-related products.

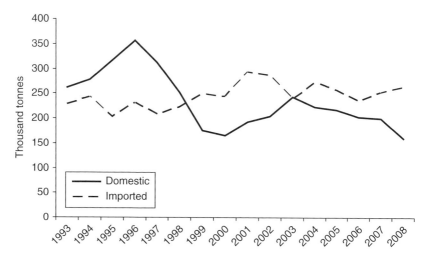

Figure 6.3   Japan's salmon supply, 1993–2008. (*Sources*: FAO, Japanese Import Statistics)

salmon is used raw in high-end markets such as *sashimi* and *sushi* at sushi bars and restaurants.

Over the past decade there has been a dramatic shift in Japanese salmon consumption from salted salmon to fresh, chilled and defrosted salmon. This shift was caused partly by increasing consumer awareness of the health benefits of reduced salt consumption, and partly by a shift in the relative shares in total salmon supply of autumn chum salmon (traditionally salted) and imported farmed salmon.

6.2.1   Japanese salmon supply

In Figure 6.3, the supply of salmon to Japan in the years 1993–2008 is divided into domestically produced and imported salmon. Total Japanese salmon supply was relatively stable during this period, after a significant increase in the late 1980s. However, the composition of supply changed significantly (Table 6.7). Three major factors drove the changes in supply. First, there were significant fluctuations in Japanese harvests of autumn chum salmon released by hatcheries in northern Japan, which accounted for 26–41% of total supply during this period. Autumn chum salmon harvests peaked in 1996 at 206 000 tonnes and then fell by almost half to 108 000 tonnes in 2000. Thus, changes in the autumn chum harvest, resulting primarily from the effects of changing ocean conditions on hatchery returns, have had a significant effect on total Japanese salmon supply and consumption. While chum is the most important species in Japan, domestic landings of pink and also sockeye have varied substantially.

Second, Japanese supply of wild sockeye salmon, most of which is imported from the USA and Russia, peaked at 132 000 tonnes in 1993 and then declined dramatically to 40 000 tonnes in 2008 (Table 6.7). Most of the decline in the supply of sockeye was due to a decline in harvests in North America. Another factor was a shift in North American sockeye salmon production from frozen to tinned product, as the relative profitability of the two product forms changed.

Third, Japanese imports of farmed salmon increased dramatically over this period. Japanese imports of farmed coho salmon (almost entirely from Chile) increased from 13 000 tonnes in 1991, peaked at 88 000 tonnes in 2001 and comprised 82 000 tonnes in 2008. Over the same time period, imports of farmed trout (primarily from Chile and Norway) rose even more dramatically, increasing from 9000 tonnes in 1991, peaking at 114 000 tonnes in 2002 (98 000 tonnes whole fish and 16 000 tonnes of fillets), and reaching about 106 000 tonnes in 2008. Imports of farmed Atlantic salmon increased from 11 000 tonnes in 1991 to 25 000 tonnes in 2008, of which 20 000 tonnes were fresh and mostly from Norway. For Atlantic salmon, the peak year was 2001 with an import of 38 000 tonnes.

The combined effect of the decline in the supply of wild autumn chum and sockeye salmon and the increase in the supply of farmed coho, trout and Atlantic salmon was a dramatic shift in the relative contributions of wild and farmed salmon to Japanese supply. Between 1991 and 2008, farmed salmon and trout increased from 15% to about 50% of total Japanese supply.

Figure 6.4 provides more detailed data for Japanese imports of whole salmon and trout for the years 1993–2008. During this period, total Japanese imports of whole salmon ranged from 202 000 tonnes in 1995 to 286 000 tonnes in 2002. Total imports have gone up and down, reflecting the combined effects of different trends in imports of different species from different countries.

The composition of Japanese imports of frozen 'red-fleshed' salmon shifted dramatically between 1993 and 2008. Over this period, frozen sockeye imports from the USA and Canada fell from 109 000 tonnes to about 25 000 tonnes, while imports of frozen Russian sockeye increased from 7000 tonnes to 16 000 tonnes. Imports of other frozen wild Pacific salmon declined from about 36 000 tonnes to 4000 tonnes. While imports of frozen wild sockeye were declining, Japanese imports of frozen farmed coho and trout increased dramatically. By the end of the period, imports of frozen coho consisted almost entirely of farmed coho from Chile. Chile also accounted for about two-thirds of frozen trout imports (including trout fillets), while Norway accounted for most of the rest.

The sharp decline in imports of wild salmon from the USA combined with rapid growth in imports of farmed salmon from Norway and Chile have led to a dramatic shift in the relative market shares of these suppliers. Between 1993 and 2006, the USA's share of Japanese imports fell from 51%

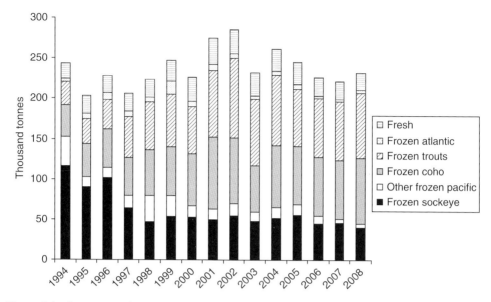

Figure 6.4   Japanese salmon and trout imports, 1994–2008. (*Source*: Japanese Import Statistics)

to less than 10%, and farmed salmon now make up more than 80% of Japanese salmon imports. This is in stark contrast to the situation in the 1980s when there was no farmed salmon and most imports were from Canada and the USA.

6.2.2   Outlook for the Japanese salmon market

In contrast to the USA and most countries in Europe, Japan may be considered a mature market for salmon. Per-capita consumption is high, salmon is widely available, and consumers are very familiar with the product. Thus it seems unlikely that total salmon consumption will grow much as a result of expanded demand. Instead, further increases in consumption will likely require lower prices, as occurred during the 1990s.

A number of other factors are likely to affect the Japanese salmon market. Japan is undergoing significant demographic transitions, including the decline of multi-generational families and the subsequent move towards nuclear families and one-person households, as well as the ageing of the population. These demographic changes have brought about changes in food consumption patterns. Sales of prepared, pre-cooked and take-out meals have increased. Older people tend to maintain a diet with more fish.

There is a growing concern for health issues. The population in general has shifted away from salty fish products to less salty ones. There is also increased awareness of genetically modified products, and people are

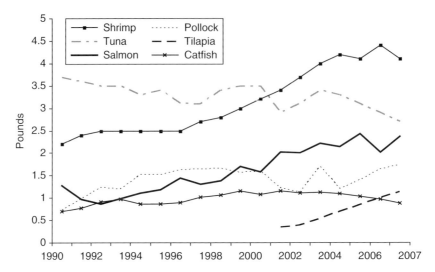

Figure 6.5   Estimated US per-capita fish consumption: top six species, 1990–2007.
(*Source*: National Marine Fisheries Institute)

willing to pay a price premium to eat 'safe' food. On the other hand, the spread of 'western' food habits has a negative impact on fish consumption.

It is clear that the Japanese salmon market will remain very large and very important. Although total salmon consumption may not rise as rapidly as in other markets, the Japanese market will provide very important opportunities for those salmon producers able to produce high-quality products that respond to the demands of Japanese consumers.

6.3   The United States salmon market

The USA is the third largest salmon market after the EU and Japan. In common with the EU but in contrast to Japan, salmon consumption in the USA is also growing rapidly. Although the USA is the world's largest producer of wild salmon, the domestic USA salmon markets are increasingly dominated by imported farmed Atlantic salmon, with Chile and Canada as the main suppliers.

6.3.1   US seafood consumption

Seafood consumption in the USA is relatively low, with an annual average per-capita consumption of 24.1 kg in 2005. However, the composition of species is very interesting. Figure 6.5 shows estimated US per-capita fish consumption for the top five five seafood species.[20] In 2007, salmon ranked

---

[20]   Please note that these numbers reflect consumed weight while the numbers at the start of this chapter and the aggregate measure reflect round fish equivalents.

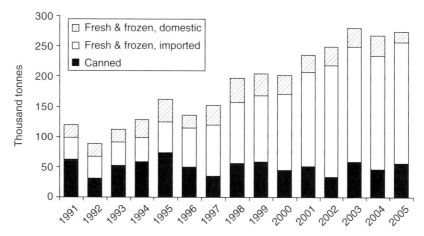

Figure 6.6   Estimated US salmon consumption (edible weight), 1991–2005. (*Source*: estimates by Gunnar Knapp based on the methodology in Knapp *et al.* (2007))

third after shrimp and tinned tuna. Estimated per-capita consumption of salmon increased rapidly from the 1990s, and reached a peak in 2005. There was a reduction in salmon consumption in 2006, while 2007 consumption was almost at the 2005 level. The reduced consumption in 2006 is most likely caused by the strong price increase, although stronger than normal negative media coverage can also have played a part. It is also worthwhile to note that four of the six most popular species, shrimp, salmon and tilapia, are primarily from aquaculture. Moreover, the consumption of these species now makes up almost 80% of US seafood consumption, while in 1985 the top six made up about 60% of consumption. This rapid growth in consumption takes place in a situation with a much more moderate growth in total per-capita seafood consumption. Also note that the consumption of the two wild species in Figure 6.5, tuna and pollock, is stagnant. Other important species in the mid 1980s like cod and flounder are moving down the list.

There are three major components of US salmon consumption: domestic fresh and frozen, imported fresh and frozen, and tinned salmon. Figure 6.6 shows estimates of total US consumption of each component in terms of edible weight. Each exhibits a different trend over time.

US consumption of domestic fresh and frozen salmon increased during the 1990s, and has been relatively stable after the turn of the century. Estimated consumption varies significantly from year to year, largely reflecting variation in wild harvests. Between 1995 and 2005, estimated consumption ranged between 18 000 and 36 000 tonnes. Because of varying catches, the species composition also varied substantially. The low-valued chum tends to be the most consumed species, although consumption of the three high-valued species – chinook, coho and sockeye – in most years is higher than consumption of the low-valued species pink and chum.

Table 6.8 US imports of fresh and frozen salmon, by product, 1990–2008 (tonnes product weight).

	1990	1995	2000	2001	2002	2003	2004	2005	2006	2007	2008
Tonnes											
Fresh fillets	0	7 044	58 125	83 182	93 638	97 893	91 719	95 455	81 49	87 813	86 24
Other fresh	16 305	42 176	56 639	63 276	71 509	63 906	59 298	70 026	81 469	84 009	84 369
Total fresh	16 305	49 221	114 765	146 459	165 148	161 799	151 017	165 481	162 96	171 822	170 609
Frozen fillets	0	1 816	9 727	14 892	22 788	25 503	28 113	29 460	35 306	31 737	27 084
Other frozen	2 887	2 742	4 828	3 786	1 400	655	3 591	2 291	2 330	2 990	2 832
Total frozen	2 887	4 558	14 555	18 677	24 187	26 158	31 704	31 752	37 636	34 727	29 916
Total	19 193	53 778	129 320	165 136	189 335	187 957	182 721	197 232	200 596	206 549	200 525
Percent											
Fresh fillets		13	45	50	49	52	50	48	41	43	43
Other fresh	85	78	44	38	38	34	32	36	41	41	42
Total fresh	85	92	89	89	87	86	83	84	81	83	85
Frozen fillets	0	3	8	9	12	14	15	15	18	15	14
Other frozen	15	5	4	2	1	0	2	1	1	1	1
Total frozen	15	8	11	11	13	14	17	16	19	17	15
Total	100	100	100	100	100	100	100	100	100	100	100

Note: Imports of fillets were not reported separately until 1995. Other products are primarily round.
Source: National Marine Fisheries Service.

US consumption of imported fresh and frozen salmon, almost entirely farmed and almost entirely Atlantic, increased dramatically from 19 000 tonnes in 1990 to 200 000 tonnes in 2008. In 2008, imported salmon accounted for more than 80% of estimated fresh and frozen salmon consumption. Imports of fresh and frozen salmon have clearly been the main factor driving the rapid increase in US per-capita salmon consumption.

US consumption of tinned salmon, almost entirely domestic production from wild harvests, is relatively stable as measured in product weight, but represents a declining share of US salmon consumption. Tinned salmon consumption also varies widely from year to year.

6.3.2 *Imported fresh and frozen salmon*

The most important factor in the growth of fresh and frozen salmon consumption has been rapid growth in imports. As shown in Table 6.8, between 1990 and 2008 total fresh and frozen salmon imports increased by a factor of 10 by product weight, increasing from 19 193 tonnes in 1990 to 200 525 tonnes in 2008. US imports of fresh and frozen salmon consist almost entirely of Atlantic salmon, virtually all farmed. Fresh salmon accounts for 75% of salmon imports, although the frozen share has been increasing gradually in recent years. A more dramatic change is the rapid growth in value-added imports. Fillets increased from 16% of total imports in 1995 to 57% of total imports in 2008 in product weight. Converted to round weight equivalents, fillets now make up over 70% of the imported quantity.

As shown in Table 6.9, Chile and Canada are the two main suppliers of imported fresh and frozen salmon to the USA, followed distantly by Norway and the UK. Chile surpassed Canada as the leading supplier in the late 1990s, and by 2001 accounted for more than half of US imports. In addition, China is becoming an increasingly more important supplier of salmon to the USA, particularly of frozen fillets.[21]

Until 1990, Norway was the leading supplier of farmed salmon. However, imports from Norway virtually disappeared from the US market after dumping allegations in 1991 resulted in a 26% import duty on fresh round salmon from Norway. Since 1999, however, there has been an increased presence of Norwegian salmon products not subject to the tariff, most notably fresh and frozen fillets, as well as frozen round Atlantic salmon. Also note the increased imports from the UK around the turn of the century and peaking in 2003, indicating the global character of the salmon market.

The rapid growth in imports from Chile started out with whole salmon but, as discussed in Chapter 3, has largely shifted to more processed products such as fillets. In 2008 Chile supplied 96 400 tonnes of salmon to the USA, of which over 75% was fresh or frozen fillets. Chilean supplies peaked at over 110 000 tonnes in 2005, but due to the current disease problems it is

---

[21] China has no domestic production of salmon and acts as a processor and trader.

Table 6.9 US imports of fresh and frozen farmed salmon, by country, 1990–2008 (tonnes product weight).

	1990	1995	2000	2001	2002	2003	2004	2005	2006	2007	2008
Chile	4 534	21 346	63 161	84 771	98 296	106 078	107 840	110 587	101 065	105 076	96 465
Canada	4 615	28 573	47 428	65 637	77 742	57 958	52 583	67 443	75 782	73 601	77 510
Norway	7 199	2 290	7 940	6 586	5 464	6 491	5 395	4 817	7 184	8 176	6 543
United Kingdom	695	814	6 583	6 266	6 048	12 531	9 077	4 873	7 712	11 688	10 603
Other countries	2 149	755	4 208	1 876	1 785	4 899	7 826	9 512	8 853	8 008	9 404
Total	19 193	53 778	129 320	165 136	189 335	187 957	182 721	197 232	200 596	206 549	200 525

Source: National Marine Fisheries Service.

unlikely that we will observe such numbers for some time to come. In contrast, fillets account for only a small share of imports from Canada. Thus Chile is the main supplier of fillets to the US market, while Canada is the main supplier of round (headed and gutted) product.

6.3.3 Market trends

Salmon is sold both in retail shops and in food service establishments (e.g. restaurants, staff cafeterias, institutional dining halls). Although no data are available on the relative shares of salmon sold by retail and food service, aggregate data for all US seafood products suggest that roughly similar shares of seafood (by volume) are sold by retail and food service. It therefore seems reasonable to assume that both retail and food service account for large shares of US salmon sales.

Salmon is the leading food service seafood menu item and is estimated to be on 39% of all food service operation menus, including 71% of 'fine dining' restaurants, 71% of 'hotel/motel' restaurants, and 49% of 'casual/theme' restaurant menus.[22] Salmon was the 'most commonly menued centre-of-the-plate fin fish' at 21% of restaurants in a 2001 survey, more than double the share of the next leading fish species (catfish and cod).[23] However, the US market also seems to be the market with most concerns about farming practices, and several organisations are claiming salmon farming to be unsustainable. There are several campaigns against farmed salmon, and farmed salmon also appear with a 'red light' on many seafood advisory wallet cards such as the Monterey Bay Aquarium's Seafood Watch. Hence, US salmon consumption still seems set to grow, but the demand development is more uncertain than in other markets.

6.4   The Russian market

While there has been a domestic fishery for salmon in Russia on its east coast, this market was not very interesting for exporters until the turn of the century. However, imports have quickly increased after 2000 and reached about 76 000 tonnes in round weight equivalents in 2008 with a total import value of €235 million. In 2006 imports of fresh salmon were reduced significantly because of import restrictions on Norwegian salmon to Russia. However, the growth in the Russian imports continued on the earlier trend after the trade issues were resolved.

---

[22]   Restaurants and Institutions 2001 Menu Census, as cited in H.M. Johnson and Associates, 2002 Annual Report on the United States Seafood Industry.

[23]   Datassential Research, Inc. research for Restaurant Hospitality Magazine, as cited in H.M. Johnson and Associates, 2002 Annual Report on the United States Seafood Industry.

Table 6.10 Russian imports of salmon (thousand tonnes product weight).

	Fresh Atlantic	Frozen Atlantic
2000	1.5	5.6
2001	2.6	10.6
2002	4.0	14.6
2003	8.8	22.4
2004	19.2	18.1
2005	31.4	29.2
2006	16.3	30.4
2007	45.6	21.9
2008	49.5	19.2

Sources: Statistics Norway, Eurostat, Chilean Export Statistics.

Table 6.10 shows an estimate of Russian imports. The estimates are based on exports from Chile, Norway and the UK (other countries export only tiny quantities, if at all). This is likely to be an underestimate, as some salmon is shipped to Russia via the Baltic countries. Russian imports are primarily whole fresh and frozen salmon. Initially, the imports were mainly frozen, but in 2004 fresh quantity was almost at the same level as frozen. In 2005, fresh was larger than frozen, before imports were almost halved in 2006. This was primarily because imports from Norway were stopped during the first half of the year after Russian food security agencies found very high levels of contaminants in some imported salmon. The problem was resolved by the Russian government certifying a limited number of Norwegian companies for exports to Russia, and the imports (and import growth) quickly reached the old level, with fresh salmon taking an increasingly dominant position. In this period, Norwegian exports of salmon to the Baltic countries increased substantially, and it is possible that re-exports from these countries at least partly circumscribed the import ban in 2006. However, salmon imports from other sources like the UK also increased. In the last few years, imports of frozen fillets have also started to reach significant volumes.

Norway is the main exporter of Atlantic salmon to Russia, with an import share of over 90%. Norway's share has been declining though, as more exporters in Chile and the UK find the market interesting as its size grows.

Originally, Russian importers demanded mostly larger salmon sizes. Most salmon were lightly salted by the importers or sold frozen to shops and markets. While this is still the most important product form, other product forms now exhibit faster growth. Only in recent years have the Russians started to heat-treat Atlantic salmon, usually by baking or frying them. New consumption patterns such as these tend to be first adopted by younger consumers with high levels of education and income. In Russia, Atlantic salmon is mostly consumed west of the Ural Mountains, mainly in Moscow, St Petersburg and other large cities with significant purchasing power. Domestic Pacific salmon is also consumed and is sold mainly frozen, salted or tinned, and dominates consumption east of the Ural Mountains.

The retail segment, particularly supermarkets and hypermarkets, is developing fast. At the end of the 1990s only a few supermarket chains had been established in Moscow, each with a few outlets. Only a few years later, large chains, both national and international, were opening up shops in Moscow and other large cities. The emergence of supermarkets and hypermarkets facilitates the distribution of salmon and increase safety and quality standards. The product range in the supermarkets is generally also wider than at the markets, and the consumer avoids the haggling on prices. In the years to come, the growth of the supermarket segment will most certainly increase the availability and sales of imported farmed salmon.

6.5   Price development

We have illustrated that the price of salmon is declining as production has increased. But what is the price of salmon? This is a surprisingly difficult question. Is it the price in Norwegian kroner, British pound sterling, Euros, US dollars or Japanese yen? Is it the nominal or the real price? Is it the producer price, the export price or the retail price? Is it the price for whole fish, fillets or smoked salmon? Is there such a thing as the price of salmon? Any measure is, of course, in its own way correct. In this section we try to provide an answer to this question, and we do that primarily by providing examples.

The main reason one can even contemplate speaking about the price of salmon is that salmon is a highly traded product. Hence, products from different producers compete with each other in different markets. If buyers perceive little difference in the product depending on its origin, or do not care or care little about where it comes from, a purchase is likely to depend only on the combination of price and quality that is offered. Similarly, a producer will primarily care about the price that is paid. Whether the product ends up in Japan or France, or as processed or unprocessed is of little concern to the farmer.

Chilean and Scottish salmon need not necessarily compete directly to have the same price development. It is sufficient that salmon from other producers compete with both parties. Hence, if Norwegian salmon is competing with Chilean salmon in Japan and Scottish salmon in France, this provides a link that will give a similar price pattern for Chilean and Scottish salmon. The relationship will be weaker the more segmented the markets become, and will break down if the markets are completely separated. This can happen in geographical space if products from different producers never meet in the same market, or in product space, as the quality of the products becomes sufficiently different.

6.5.1 Real versus nominal price

Inflation measures how much prices have increased during a period for a given basket of goods. This means that inflation reduces the value of money in terms of what can be purchased. The price of a product as given by the price tag is known as the current price or the nominal price. Over time money is losing value due to inflation. Hence, to compare prices over time, one must adjust for the inflation or, as the process is also known, deflate the nominal price. By removing the inflation one gets the real price. If the real price is constant over time, the nominal price will be increasing at the same rate as the inflation.

To adjust for inflation is important when one is interested in how the price of a product develops over time. Hence, it is important to use real prices when one is asking questions like 'Is salmon cheaper today than 10 years ago?' If the real price is decreasing, a buyer will also find that the price has become cheaper relative to an average of all goods, while if the real price is increasing, the product becomes relatively more expensive.[24] However, if one is interested in comparing the development in the prices of different products over time, it matters less whether one uses real or nominal prices.

How important is inflation? That depends on the rate of inflation and the length of the period of comparison. In Figure 6.7 we show the nominal and real export prices for Norwegian salmon, with 2008 as the base year. Inflation is measured by a price index, normally the consumer price index (CPI). Hence, when the price is measured in Norwegian kroner, we adjust for inflation or deflate the price using the Norwegian CPI. We see that the real price is much higher than the nominal price in 1981. In fact at NOK88.79/kg the real price is about two and half times as high as the nominal price at NOK32.74/kg. In the early 1980s, the real price was stable while the nominal price was increasing, and from 1985 the real price was declining much faster than the nominal price. Undoubtedly, there has been a decline also in the nominal price during this period as it moved from NOK32.74/kg in 1981 up to NOK46.74/kg in 1985, then down to NOK21.73/kg in 2003 and up again to NOK31.81/kg in the peak year 2006 before ending up at NOK26.95/kg in 2008. However, this is not much relative to the real price, which moved from NOK88.79/kg in 1981 up to NOK93.38/kg in 1985 and then down to NOK23.72/kg in 2005 before increasing to NOK26.95/kg in 2008.

---

[24] It is of interest to note that the real prices of all large-volume foodstuffs (grain, poultry, etc.) have been decreasing, despite some short-run cycles like the food price crisis in 2007–2008. Moreover, in general the wealthier a country, the lower the share of disposable income that is used on food.

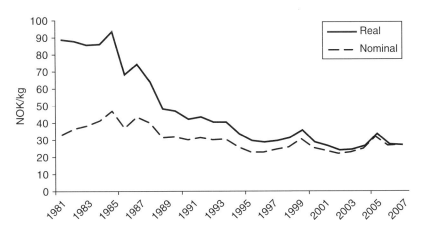

Figure 6.7   Norwegian real and nominal export price, 1981–2008 (2008 = 1). (*Source*: Norwegian Seafood Exports Council)

6.5.2   Exchange rates

The exchange rate gives the price of one currency in terms of another currency. Hence, the exchange rate between Norwegian kroner and Euros (NOK/€) reflects how many kroner one has to pay for €1. Exchange rates are often quite volatile. There are several reasons for this. For instance, if inflation is higher in the UK than in the USA, one would expect that the pound is weakening against the dollar since the pound over time will buy relatively less than a dollar. Other factors like interest rates, economic policy and economic development also influence the exchange rate. Surely then, there cannot be a single global price for salmon? We are here interested in at least two questions. Have buyers of salmon in all markets experienced the same decline in prices as Norwegian producers? And are the prices of salmon similar in different markets if they are converted into the same currency?

To compare how the real salmon price develops over time, we must first convert the nominal price into the currencies of interest, and then adjust for inflation in the different countries. Furthermore, since we are interested in the changes in the prices, we divide all the prices by the price in a base year, so that they are normalised to one in that year. The vertical axis then measures the relative change in the price. In Figure 6.8 we show the Norwegian export price in Norwegian kroner adjusted with the Norwegian CPI, the Norwegian export price converted to US dollars and adjusted with the US CPI, and similarly the price in Euro and British pounds, all normalised to one in 1985. The most important insight from this figure is that the salmon price has declined substantially in all currencies, and the main trend is similar. Looking at 2008 prices, the price in Norwegian kroner and Euros has declined slightly more as the prices in 2008 were about 29% of the 1985

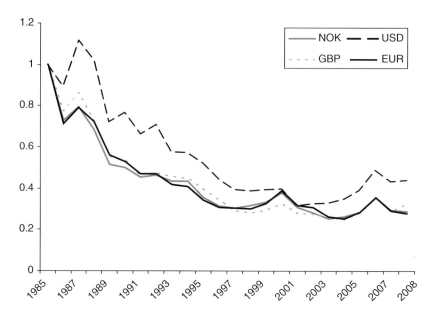

Figure 6.8 Real salmon price in different currencies, 1985–2008. (*Sources*: Norwegian Seafood Exports Council and authors' computations)

level, compared with about 44% in US dollars and 32% in British pounds. There are also some differences in the development in that the price is relatively higher in the USA in the late 1980s. The development is very similar in the 1990s before the price in Euros starts to increase relative to the others just before the turn of the century, and the US price follows suit in 2003 and 2004. Hence, although the main trend is the same, how much the real price of salmon has declined for consumers in the different regions at different points in time varies with the development in exchange rates and inflation.

6.5.3 Prices in different markets

Currency markets are very efficient, so whether a Norwegian exporter sells in Japanese yen or Euros does not matter as long as the price is the same in Norwegian kroner. This is indeed what one observes, as shown in Figure 6.9, where the Norwegian export price for whole salmon to France and Japan is shown. There is a difference in the price levels, primarily due to additional costs involved with the airfreight to Japan. There is also some short-run variation, indicating that the prices are not perfectly correlated. This is because the exporter, although keeping track of price developments in both markets, cannot react instantaneously to changes. Moreover, there will also be some variation because quality composition differs over time.

In Figure 6.10, we show import prices for Norwegian and Scottish whole salmon to France in Euros per kilogram. Again, the prices follow each other very closely, although with some short-run variation. There is an exception

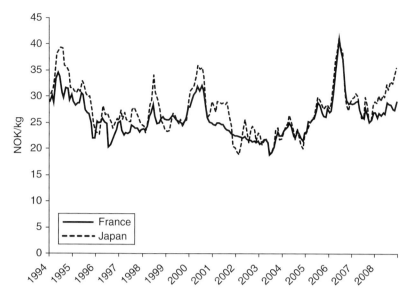

Figure 6.9   Norwegian export prices to France and Japan, 1994–2008. (*Source*: Norwegian Seafood Exports Council)

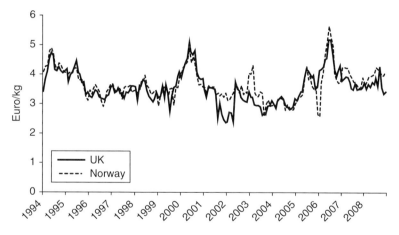

Figure 6.10   Import prices for fresh salmon to France, 1994–2008. (*Source*: Eurostat)

in late 2002 to early 2003. This was caused by a minimum import price for Norwegian salmon that was strongly binding in that period.[25] The mesage from this figure is that many buyers of salmon in France do not care whether

---

[25]   In a substantial part of the period 1994–2006, there were minimum import prices for Norwegian salmon to the EU. However, as indicated in the figure, these minimum import prices were not strongly binding most of the time.

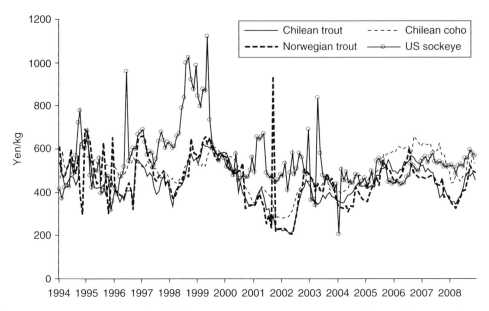

Figure 6.11   Import prices for fresh salmon to Japan, 1994–2008. (*Source*: Japanese Import Statistics)

the salmon comes from Norway or Scotland. What matters is the combination of price and quality.

A similar story is told in Figure 6.11, where we show import prices of Chilean salmon trout, Chilean coho, Norwegian salmon trout and US sockeye to Japan for the period 1994–2008. The correlation between the prices for the three farmed species is higher than for any of them relative to wild sockeye. The price of sockeye increases somewhat relative to the other three prices in the salmon year 1997–1998, and substantially in 1998–1999, before it falls back in line in the second half of 1999. This is primarily due to weak seasons with a low harvest of wild sockeye. Hence, one can say that sockeye is part of the Japanese salmon market, but in years with a limited supply sockeye will form a separate market segment. This indicates that some Japanese consumers strongly prefer sockeye to the other species and are willing to pay for it. However, these consumers have to pay extra to obtain sockeye only in years with a low supply, because in normal years there are a number of consumers willing to shift between sockeye and the farmed species. Relative to the other short-run deviations in the different prices we have seen, the deviation for sockeye is large. Since the price is back in line with the other prices after one or two seasons, this is an indication that sockeye is a part of the salmon market and does not form a separate segment.

Comparing prices is known as market integration tests or market delineation tests. This is discussed in greater detail in Chapter 7. The reader will benefit further from the exposition here after having read that chapter.

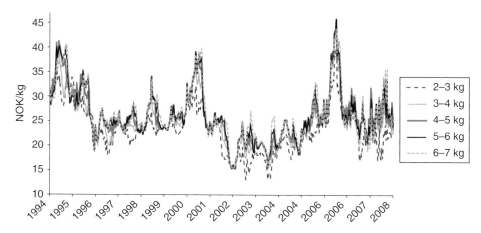

Figure 6.12 Norwegian producer prices for different weight classes, 1994–2008. (*Sources*: Norwegian Seafood Federation, NOS Clearing)

6.5.4 Prices in different weight classes and for different product forms

Salmon is not a homogeneous product. There are substantial differences when it comes to quality, which can be assigned to different attributes such as colour, fat content and size. In different markets, one may be willing to pay a premium for a specific attribute and this may cause segmentation. On the other hand, it is costly to provide special attributes, and one would expect a farmer to supply the product only if the higher price provides sufficient compensation. Hence, the cost will limit the supply of specific attributes and tend to keep prices in line with each other.

One of the most important quality attributes is size. The uses of salmon for several purposes depend on size. There is therefore a tendency that larger salmon obtain a higher price per kilogram. This is seen in Figure 6.12, which shows Norwegian producer prices by weight class. What is most interesting in the figure is that the dispersion in the prices seems much larger than, for example, the import prices from Norway and Scotland to France as shown in Figure 6.10. The potential for segmentation is accordingly even larger here. Moreover, there seems to be patterns in the price development.

This is indeed the case, as shown in Figure 6.13. Here, prices for the other weight classes are divided by the price of salmon that weigh 3–5 kg. Hence, the figure shows the price development for the different weight classes relative to the price of salmon weighing 3–5 kg. Salmon of this weight have been chosen as a base since this is the expected weight of a salmon that has been in the sea for 1 year. The relative prices show a surprisingly stable pattern. The 1–2 kg weight class is most volatile. In the winter months it has the same price level as the 3–5 kg fish, while in the summer months the price is much lower, as low as 70% of the price of 3–5 kg fish. Fish of 2–3 kg

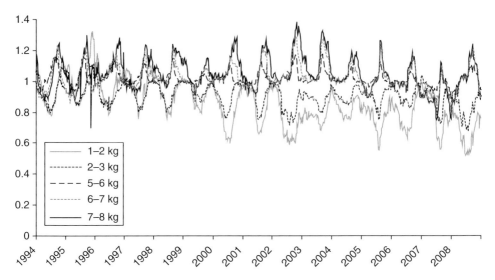

Figure 6.13 Relative prices for different weight classes, 1994–2008. (*Sources*: Norwegian Seafood Federation, NOS Clearing, authors' computations)

have a similar but weaker pattern. Similarly, the large fish of 7–8 kg fetch a much higher price in autumn, with a premium that can be as high as 20%. The 6–7 kg fish show a similar but weaker pattern.

Asche and Guttormsen (2001) showed that the main reason for the patterns in the relative prices is the biology of the salmon. Most salmon are transferred to the sea in May and, as shown in Chapter 2, the growth is uneven over the year. Hence, salmon will tend to grow in cohorts. There will be relatively moderate quantities of small salmon in the winter months because salmon transferred to the sea in autumn are not large enough to sell, and those that were transferred to sea the previous spring are already larger. In this period the price for the small salmon is therefore comparable to that of the larger salmon. However, in the summer months when the small salmon are abundant, it fetches a relatively low price. Similarly, in order to avoid large salmon becoming sexually mature in autumn, most are sold during summer. The availability is therefore limited during autumn, and therefore large fish fetch a relatively high price in this period. Since the pattern is predictable, the farmers adjust to it to the extent that it is profitable, and one cannot really speak about different market segments for salmon of different sizes.

It is of interest to note the similarity in the explanation above with why the sockeye price deviated from the other import prices to Japan, and why there is a positive premium for large salmon in one period of the year and a negative premium for small salmon in another period of the year. The key factor is availability, and when availability becomes sufficiently restricted,

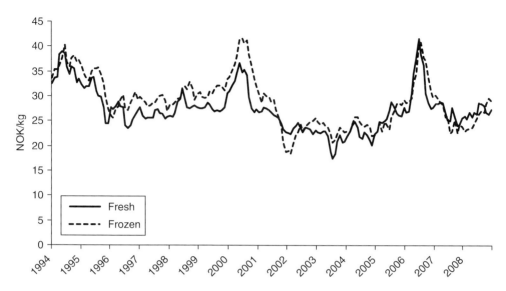

Figure 6.14   Norwegian export prices of different product forms, 1994–2008. (*Source*: Norwegian Seafood Exports Council)

there seems to be a group of consumers willing to pay a premium for a specific characteristic.

There is also a link between different product forms of salmon, since they are all using the same raw material. Certainly, processing costs must be added to the price to reflect the extra value added, but movements in the salmon price will then also influence the price of the processed products. However, the link becomes weaker, the more processed the product becomes. This is only natural, since the share of the total cost that salmon makes up is then reduced. In Figure 6.14, we show Norwegian export prices of fresh and frozen salmon. As frozen salmon is a type of processing that adds relatively little value, we see that the prices move fairly closely together.

There is also a link between the prices at different levels in the supply chain. However, one would generally not expect this link to be equally strong, since the difference in the prices at the different levels, the margin, is largely determined by other costs. In Figure 6.15 we show the Norwegian export price and the French supermarket retail price for whole salmon. As one can see, both prices have the same weak decreasing trend (which would have been much stronger if we adjusted for inflation), although the short-run volatility differs substantially. The volatility at the retail level is partly caused by salmon being used in promotion, and then sold at lower prices. As shown in Figure 6.16, the margin over time is almost constant although with a slightly decreasing trend. This is an indication

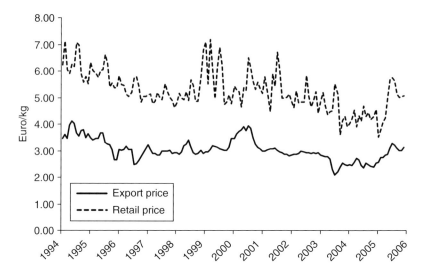

Figure 6.15   Norwegian export prices and French retail price, 1994–2006. (*Sources*: Norwegian Seafood Exports Council, Oifremer)

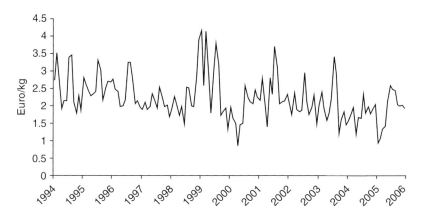

Figure 6.16   Margin between the Norwegian export price and French retail price, 1994–2006. (*Sources*: Norwegian Seafood Exports Council, Oifremer, authors' computations)

that there are productivity improvements also in the distribution channel between the exporter and the retailer.

Bibliography

Shaw and Muir (1987) and Bjørndal (1990) are good early overviews of the salmon market. Anderson (2003) provides a good overview of the seafood market in general where salmon holds an important role since it is among the more dynamic sectors. FAO, including Globefish, is a very good source for statistics and market analyses.

The difference between real and nominal prices is discussed in most introductory texts in economics. Factors determining exchange rate movements are a topic in many texts in international economics and macroeconomics. One good example is Krugman and Obstfeldt (1994). The impact of exchange rates on international trade with respect to market integration is also widely discussed in the international trade literature. Goldberg and Knetter (1997) is a good introduction. Asche and Tveterås (2008) and Xie *et al.* (2008) provide analyses of Japanese salmon imports from Chile, Norway and the USA. Roheim (2009) gives an insight into seafood guides.

A number of studies have investigated relationships between salmon prices. Asche and Sebulonsen (1998) investigate the relationship between Norwegian and Scottish salmon in France. Asche *et al.* (1999b) show that farmed and wild salmon compete in the same market, as for fresh and frozen. Asche (2001) shows that prices for fresh salmon in Europe, Japan and the USA are highly correlated. Asche *et al.* (2005) investigate the Japanese salmon market focusing on different red-fleshed species from different origins.

Asche and Guttormsen (2001) provide an analysis of the prices for different weight classes of salmon. Asche *et al.* (2001a) show that because the patterns in the prices are deterministic, the markets are not sufficiently segmented to prevent the generic term 'salmon' from being a sensible category. Oglend and Sikveland (2008) look at salmon price volatility.

Guillotreau *et al.* (2005) look at how prices in the supply chain for salmon in France are related, focusing on the dynamics of the adjustment process. Asche *et al.* (2007c) provide an analysis of how the prices in the supply chains for Norwegian and Scottish salmon are linked, while Asche *et al.* (2007b) compare the supply chain for salmon with the supply chain for cod. Kvaløy and Tveterås (2008) look at the vertical structure in the Norwegain supply chain.

7 Competitiveness and Market Structure

Different salmon markets have different product preferences, and distribution channels differ within, as well as between, markets. Over time, these channels have been changing for a number of reasons, including cultural and historical differences and variations in market structure. This chapter considers some of the most important features of the salmon market with respect to competitiveness and market size, and how they have changed over time. This will also provide some important insights into factors determining prices.

While many aspects of market structure are important in determining price, two are particularly important. The first is to what degree different products compete with each other, or how substitutable they are. The second is to what extent a firm or group of firms can exploit market power: are they able to influence prices and profits by adjusting quantity supplied? In Chapter 4, productivity growth was identified as an important factor in explaining production growth. Market or demand growth is a similar factor on the market side. There is no doubt that the salmon market has grown substantially and producer, consumer and supply chain characteristics have changed as volume has increased.

Hopefully these issues are addressed in an accessible manner that does not require previous knowledge of economics. However, the issues are at the core of how markets operate, and therefore an appendix laying out a simple economic model of a market is provided for interested readers who do not have a background in economics.

7.1 What is a market?

The first important question is 'What is the market?' The simplest definitions are often based on the relationship between prices. Stigler (1969) defined a geographical market as 'the area within which the prices tend to uniformity, allowances being made for transportation costs'. This definition is easily extended to allow for price differences due to quality variations. Hence, it can be said that a group of products or regional markets constitute one single market, or are fully integrated, if prices move proportionally with respect to each other over time. For instance, the markets for Chilean salmon in Miami and Chicago are fully integrated even though the price is higher in Chicago, if the price difference is only due to higher transportation costs to Chicago. This is often known as the Law of One Price.

With respect to consumer preferences, two products are perfectly substitutable if the consumer does not perceive any difference between them. Similarly, a producer can be indifferent between producing two goods, say fresh and frozen salmon. The producer will produce the product form that is most profitable, and both will be available in the market, provided they give the producer the same profit. Accordingly, market integration can be driven by consumers as well as by producers, and it is sufficient that one of the parties perceives the market as integrated. For instance, Norwegian fresh salmon sold in France does not compete directly in retail sales with Norwegian fresh salmon sold in Japan. However, the Norwegian farmer will sell the salmon to the buyers who offer the highest prices, so for the farmer there is a close link between the markets. If demand in France increases, French buyers will bid up the price, and Japanese buyers will then have to adjust their price accordingly to get salmon. However, as their demand has not been shifting, they will buy less, shifting quantities from Japan to France. In this way the market is integrated, as indicated by the close relationship between prices shown in Figure 6.9.

If different products can be substituted only partly, the goods are imperfect substitutes. Two markets are related or partly integrated if they influence each other to some extent, but not completely. An example could be a French salmon smoker who prefers fresh farmed salmon as an input. If the price of fresh salmon increases a little, the smoker may continue to buy only fresh salmon. However, if the price increases substantially, he may start purchasing cheaper frozen Pacific salmon as well. In this case, frozen Pacific salmon is not a perfect substitute for fresh farmed salmon, as the smoker does not shift his demand immediately in response to the price increase. Rather, it is an imperfect substitute since the smoker shifts reluctantly and only when the price difference becomes sufficiently large. Viewed from a price perspective, as the price of one good increases, the prices of goods that are imperfectly substitutable will tend to follow in the same direction, but not in the same magnitude. The more substitutable the products are, the more closely the prices move.

Two goods do not compete if they are not substitutable at all to buyers or producers. In this case, if the price of one product changes, it will have no impact on the price of the other, and the products do not compete in the same market. For instance, increased demand and prices for DVDs will not influence the demand for or supply of salmon.

7.1.1 Market size

One can think of market size as the quantity of a product consumers are willing to buy at various prices, or how much the consumers are willing to spend on a product at different prices. This is normally represented by a demand schedule; there is a negative relationship between quantity demanded and price so that the demand schedule is downward sloping. For any product one can increase the quantity sold by reducing the price. However, if the market size increases, the producer will receive a higher price for all quantities supplied, because the demand schedule shifts outwards. That is, for all prices, the buyers will purchase more of the product. Similarly, if market size is reduced, suppliers will receive a lower price for any quantity supplied. Hence, changes in market size are as important on the demand side as productivity changes are on the production side.

Several factors can influence market size. The most obvious are economic, such as prices of competing products and income. If the price of a competing good is reduced, this product becomes more attractive for the buyers. They will therefore purchase more of this cheaper product, and reduce the demand and thereby the market size for the product of interest. Income growth makes people better off, and will increase demand for most products. The impact of income growth is different for different products, and income-driven market growth will tend to be less for staple foods than for luxury products. For traded goods, like salmon, exchange rates are also very important because they influence the relative buying power between countries. For instance, when a currency appreciates (i.e. it costs less to buy a foreign currency), the buying power of the people in the country with the appreciating currency increases. For producers exporting to this country, this will lead to an increase in import demand and market size, as the buyers find that imports have become cheaper because the price in their currency has been reduced. If the exporters' currency appreciates, prices in the importers' currency increase, and demand is reduced. The exporters will then experience a negative shift in market size.

Marketing is of special interest since this is the means firms use to try to influence market size directly. The main objective of marketing is to increase the number of consumers buying a product and to persuade existing consumers to maintain their purchases or buy even more. This will lead to an increase in market size. Marketing is a controlled positive 'shock' to the market. Population growth also expands market size. Other shocks can also lead to greater consumer interest in a product. For instance, increased awareness of the positive health benefits of eating fish is likely to increase demand for salmon. Negative shocks tend to have a similar form. For example, newspaper stories about negative environmental impacts of salmon farming can lead to a reduction in demand.

7.1.2 Market power

Market power exists when an economic agent by his purchasing or selling behaviour can influence price in a profitable way. When an agent takes such actions, he exploits market power. There are two main forms of market power. The most common is when a seller of a product exploits market power. In this case the quantity sold will be lower than in a competitive market, and the price higher. This is known as monopoly when there is a single firm and oligopoly when there are a few firms that can jointly exploit market power. Recently, purchasing power has also become a concern because of the increased concentration in the retail sector. In this case the buyer with market power purchases a lower quantity and with this action forces the supplier to accept a lower price. This is highly relevant for seafood, as the supermarket chains' share of retail sales now is more than 80% in several countries (see Murray & Fofana 2002 for an accessible account for the UK). Here the focus will be on seller power, as the discussion is easier to follow. However, buyer power works in a similar way.

For a firm to exploit market power, there must be limited competition. The reason for this is straightforward. We saw in Chapter 4 that high profits are the market's signal that it wants more of a product, but as a higher quantity is supplied, the profit margin is reduced as prices fall. When a firm exploits market power, it becomes more profitable because it is able to increase margins by restricting supply and thereby raise the price.[26] However, this makes the market very attractive for others to enter. If others can enter they will erode margins by increasing supply. Hence, competition will erode market power and for someone to exploit market power, there must be something to prevent or at least limit market entry.

A condition for exploiting market power is relatively few agents in the market. With many agents, there will always be competition. Since there are many salmon producers, exploitation of market power at the firm level has traditionally not been a concern. However, with increasing concentration in the industry and the growth of multinational companies, competition authorities get involved when the largest companies try to merge. Moreover, the firm level is not necessarily the relevant level of analysis. Most of the salmon is produced in a few countries. If countries are the relevant level of analysis, it can certainly be claimed that there is potentially weak competition in the salmon market. The fact that the market also seems to be regionalised, with Norway and Scotland the dominant suppliers to the European market and Canada and Chile to the US market, will further accentuate this problem if there is little competition between these markets. If countries are the relevant level of analysis and there is a limit on suitable

---

[26] This relationship can sometimes be made the other way in that the firm that exploits market power sets the price, and thereby reduces the quantity demanded.

locations for salmon aquaculture, this is an entry barrier. On the other hand, with the strong increase in production, it is difficult to argue that the quantity supplied has been limited.

Retailers or firms at other levels in the supply chain can also exploit market power by limiting quantity to fetch a higher price. In and of itself, this would then limit development of the salmon industry. Again, the strong increase in salmon supply can be used as an argument against supply limiting as a serious issue. However, it is certainly possible that margins on salmon from some retailers have been high.

Market power for producers is interesting in terms of timing and trade. In industries where there is a long time between the production decision being made and the product being ready for market, there is an opportunity for exploiting market power in the short run because competitors are unable to respond. The salmon market structure makes this possible because of the biological production process. There is a significant delay after the decision to produce a salmon is taken until it is marketable. However, in the long run, an agent that limits supply and achieves a higher than normal margin will lose market share as competitors respond by increasing their supply.

In most trade conflicts in the salmon market, measures are directed at one or a few producing countries, usually Norway and Chile, rather than all producers. Trade measures could be expected to make suppliers uncompetitive as producers who do not have these extra costs would increase supply and take over the market. This did indeed happen with Norwegian salmon in the USA, as discussed later in this chapter. However, it did not happen with Chile in the USA or Norway in the EU. This is a sign that Chile and Norway may have some market power, if and when their producers are coordinated. It can happen with regulations, such as the Norwegian feed quotas, and it can also happen with the trade measures themselves, as they limit supply and thereby increase price in the importing country. The most important issue in this context is that if the trade restrictions are to work, the producers that have their market access restricted must have some market power. If not, other producers will take over their market share.

7.2 The salmon market

Over the years many have asked 'What is the market for salmon?' Is there a global market or is it segmented into regions (Europe, America, Asia) or product forms (fresh and frozen)? The answer depends on different factors, such as when the question was asked and the level in the supply chain being considered. This section tries to explain what the salmon market is and how it has developed by answering the question at different points in time and in different places.

7.2.1 Pacific salmon

Around 1980, when salmon aquaculture was in its infancy, there was a substantial market for Pacific salmon based on annual wild landings of about 560 000 tonnes. This market had three main segments: two storable products, tinned and frozen salmon, and fresh salmon sold in the harvesting season relatively close to the landing ports.[27] The segments were further divided by species, and not all Pacific species were sold in all segments. The fresh segment was the most valuable and the prime species, chinook and coho, were sold primarily on the Canadian and US west coasts. The geographical limitation of the market was primarily due to low availability and lack of logistics.

The canning process requires a substantial investment in processing equipment, and in other input factors such as tins. As these investments have been made before the canneries decide how much fish to buy during each season, their variable costs are relatively small. Within the price ranges observed, the canneries would therefore buy enough fish to exploit their capacity. As the canning cost was substantial and few consumers substitute tinned salmon for fresh or frozen product, this market constitutes a separate segment after the fish has been tinned. However, at the point of landing the fishermen would tend to favour the buyer who offers the best price. Hence, at this stage the processors of frozen and tinned salmon compete. The sunk costs involved in the canning process limits the competition within a year, but as the landing price influences the profitability of the canneries, the landing price one year will influence the number of tins purchased the next year and, together with expected prices, influence quantity produced.

Within a given season the residual of the wild landings would tend to be frozen. Most of this would go to Japan, but substantial quantities would also be consumed in North America and exported to Europe. However, some market segments were high paying and in the long run would compete with canners for the salmon. In general though, the cost of freezing salmon is much lower than canning them. There is a strong tendency for frozen (and fresh) prices to reflect the landing price. For tinned salmon there is a much weaker link. Frozen is also the only product form that utilises all the Pacific species, as prime-quality chinook, coho and sockeye as well as low-valued pink and chum are shipped frozen.

Wild Atlantic salmon was and is available in limited quantities along the east coast of North America and in northern Europe. It is a seasonal product fresh, but has a longer season as smoked and frozen. In any form it is regarded as a luxury product and commands a very high price and, as such, largely constitutes a separate market segment.

---

[27] There was also a fourth segment. The Soviet Union landed substantial quantities of Pacific salmon. However, the fish were sold domestically under the central planning system and were not relevant to the rest of the world markets.

7.2.2 The early development of the Atlantic salmon market

Farming of Atlantic salmon was pioneered in Norway in the late 1970s. The natural outlets were the markets where wild salmon was sold, i.e. high-end restaurants and gourmet shops. Salmon was sold either as fresh or high-quality smoked. Farmers in other countries followed similar practices. Sales tended to follow the seasonal patterns established by wild salmon, with a particularly strong season around Christmas. This suited the farmers well, as the biological growth cycle to a large extent determined when production cost was lowest. However, from the early days regularity of supply was an important competitive advantage, particularly with respect to high-end restaurants. As farming practices improved and production increased, the season length increased and an Easter season was created. Still, the increased production had a limited price effect, as it was primarily satisfying latent demand that could not be satisfied by the seasonally available wild salmon. As can be seen from Figure 1.2 in Chapter 1, this appeared to be the case until 1985, as prices were stable or increasing. Only after 1985 did the price start to decline.

As one market was saturated and pressure on prices commenced, new markets were sought. There are substantial economies of scale in transport and logistics, and accordingly producers tended to target one geographical market at a time. The first target was France, the largest seafood importer in Europe, with one of the largest high-end markets. Moreover, it takes only 24 hours or so to transport salmon from Scotland or western Norway to Paris by road. It was possible to guarantee delivery of a fresh product that would reach the market less than 3 days after the fish came out of the sea.

As the geographical area where the product was sold expanded, a number of innovations were made with respect to logistics, preservation and packaging. The development of leak-proof styrofoam packing made airfreight feasible. In the mid 1980s the trade flow from Norway took a surprising turn as the USA, due to the use of airfreight, became the largest export market after France. Exports to the USA increased from virtually zero in 1981 to 9600 tonnes in 1986 (almost the same as France). The US east coast was a market with a tradition for consuming salmon, but where only very limited quantities of wild Atlantic salmon were available. Further, very little (and virtually no fresh) high-quality Pacific salmon was sent to this market. Finally, it was a market with substantial purchasing power, willing to pay high prices to cover high transportation costs.

The use of airfreight was important as it removed the barrier that distance had previously presented to a global market for fresh salmon. Post airfreight, the market developed in two significant dimensions as production has increased. The geographical size of the market expanded as it became possible to reach virtually any place in the world with salmon by airfreight. In 2008, Norway exported fresh salmon to more than 100 countries. It also allowed producers in any location to access the market. This has enabled Chile to become the world's second largest salmon producer. The other development has been expanded supply to markets where salmon can be carried cheaply by road, allowing new sales outlets and product forms to be developed. Norwegian and Scottish salmon can now be found all over Europe, while Chilean salmon is widely available in South America.

7.2.3 Early development of the farmed Pacific salmon and salmon trout market

As the production of farmed salmon expanded substantially in the 1980s, producers tended to target geographically close markets. Atlantic salmon was produced on the Canadian and US east coast, while Pacific salmon (coho and chinook) was produced on the west coast. In Chile, coho was the favoured species early on, together with salmon trout, while chinook was the preferred species in New Zealand (although salmon is not native to the southern hemisphere). However, Japan, the main export market for these producers, was accustomed to consuming frozen Pacific salmon that was thawed and salted. Hence, farmed Pacific salmon produced for consumption outside North America was primarily exported as frozen, and in direct competition with landings of wild salmon.

Trout farming has a longer tradition than salmon farming and the traditional production technology has been freshwater farming in small operations. However, as salmon farming technology improved, some trout operations, primarily in Scandinavia and the UK, also took on this technology. Large trout, or salmon trout, are in many ways similar to the Pacific species, with a more deeply red flesh than Atlantic salmon. A much higher tariff on trout in comparison to salmon prevented it from being exported to the then European Community (EC). In the mid 1980s, Finland was the leading producer and exporter of trout (see Chapter 3). However, in the early 1990s production increased substantially in Chile and Norway while targeting the Japanese market, giving these countries the lead as the dominant actors in this market. The main market for salmon trout in Japan was similar to that for wild salmon, and most of the fish were shipped in frozen form.[28]

Demand for seafood, including imported seafood, weakened substantially in Japan after 2000. This has caused production to level off in Chile and be reduced in Norway (as discussed in Chapter 3). However, recently, a new market for frozen salmon trout has developed in Russia.

7.2.4 Fresh versus frozen

In the 1980s, there was discussion as to the best way of shipping salmon. Until then, producers had the most experience with frozen product. Moreover, a frozen product was clearly the cheapest to transport, and before airfreight

---

[28]   The smaller trout farmed in continental Europe are sold as portion-sized fish and are a separate market from salmon and large trout.

became a possibility it was the only way to reach many markets. However, producers found that the main competitive advantage for farmed salmon was the ability to supply the fresh product with a high degree of reliability. This might be expected since for most fish species, fresh is the most sought-after product form and therefore provides the highest price to the producer. However, new markets are still created by first shipping frozen salmon (because of its storability) before quantities become large enough to justify investment in the logistics necessary for fresh supply chains. The markets in Russia and other eastern European countries are recent examples of this practice.

As production of Atlantic salmon increased, a very high share continued to be sold as fresh product. Until the early 1990s a substantial proportion, up to 30%, was also sold as frozen, as European processors carried some inventory and producers tried to enter the Japanese market. However, as logistics improved, processors stopped maintaining inventories, and in the Japanese market the pinkish colour of frozen Atlantic salmon could not compete in the frozen/salted market with the redder Pacific species and salmon trout. On the other hand, the market for higher-priced fresh air-borne Atlantic salmon continued to increase.

In Chile, large quantities of coho, salmon trout and Atlantic salmon were produced. Until the early 1990s, substantial quantities of frozen coho, as well as some fresh, were exported to the USA where there was a traditional market for coho. Considerable quantities of frozen Atlantic salmon were exported to Japan. However, it was realised that the US east coast (where most of the US-bound exports went) had little taste for frozen salmon but a strong preference for fresh product, while the Japanese did not care much for frozen Atlantic salmon. From then on, Chilean coho and salmon trout have almost exclusively been shipped frozen to Japan, while the Atlantic salmon primarily goes fresh to the USA. Atlantic salmon's share of Chilean production has since increased substantially.

Since the mid 1990s virtually all farmed Atlantic salmon produced is sold as fresh product, while almost all farmed Pacific salmon and salmon trout are sold as frozen.[29] Moreover, it is primarily Atlantic salmon that expand the market and from which new product forms are derived. Farmed Pacific salmon and salmon trout have for the most part been sold into the traditional Japanese market where they are thawed and salted. However, after the turn of the century sales of frozen salmon have increased. They are primarily sold to processors as an input in highly processed products and to catering companies.

7.2.5   Declining prices, new sales outlets and product forms

The high-end market segments that were initially targeted for farmed salmon were relatively small. The first signs of market saturation were seen in the mid 1980s, as prices started to decline after 1985 (see Figure 1.2).

---

[29]   The main exception is low-quality fish to markets with a lower willingness to pay.

Prices had to be reduced to attract buyers for the still increasing supply. In addition, to expand the market exporters and processors had to start looking for new product forms.

The first candidate was the relatively large market for smoked salmon in Europe. More than half of the 20 000 tonnes of frozen Pacific salmon imported to Europe was used as an input by smokehouses, primarily in France but also in other countries. Although smoked Atlantic salmon was sold in the high end of this market, Pacific salmon provided the largest quantities in the early 1980s. However, as prices of farmed salmon started to decline, smokers increasingly used farmed Atlantic salmon as an input. This was a hugely successful move, as consumers perceive smoked Atlantic salmon to be of higher quality than smoked Pacific. The declining price of farmed salmon after 1985 allowed the retail price of smoked to decline and thereby increase sales. At the same time, the price decline was much slower than what would have been experienced without this additional market to absorb a substantial part of the increased production. The smoked salmon market quickly became very important, and by the early 1990s as much as 50% of salmon consumed in Europe was smoked, with continued increases in production. Smoked salmon is still an important product form, and in many European markets it makes up as much as one-third of total consumption.[30]

As prices for farmed Atlantic salmon continued to decline, an increasing number of products were produced. However, with the exception of smoked salmon, virtually all were fresh. For instance, in France, in 1990 as much as 90% was sold as whole salmon to the consumer, while in 2000 the share of filleted and other prepacked products had increased to over 70%. In addition, salmon became increasingly popular in more value-added products, and there is currently a rapid expansion in the number of product forms available. This expansion has also caused the frozen salmon market share to start increasing again due to its use as an input in highly processed products.

Chilean producers made a major step forward in the early 1990s with the introduction of the pin-bone-out fillet. Until then, the US farmed salmon market had primarily been a market along the eastern seaboard where whole salmon was presented on seafood counters. With pin-bone-out fillets, the Chileans opened a completely new market in the Midwest, and attracted people who previously barely ate fish at all to consume substantial quantities. In recent years, Chilean exports of frozen salmon have also picked up, targeting market segments where salmon are used as an ingredient in highly processed products as well as discount stores.

How much salmon is going to be sold at profitable prices will be determined as much by market growth as by productivity growth. To a large

---

[30] This is based on raw weight, i.e. one-third of the salmon as measured in whole fish equivalents is used for smoking.

extent this will depend on firms' ability to create new markets. As salmon is sold in most countries in the world, expanding geographical markets is not really an option any more. The product can be made affordable to consumers who could not previously buy it by lower production and distribution costs, or new product forms can be created so that existing markets become deeper. Making salmon more affordable is a strategy that will work in some markets. Most people outside the EU, Japan and the USA cannot afford salmon today, even though prices have declined rapidly. Income growth that makes salmon affordable will also help expansion. This seems to be a main driver behind the substantial increase in consumption in Russia and eastern Europe during the last few years.

The largest potential may still be in further product development. Stable supplies of fresh salmon at relatively low prices have, as noted above, given rise to an increasingly large industry producing value-added products. If this process continues, there is a substantial potential for increased market growth. In particular, in the most advanced markets prepacked salmon are increasingly presented on shop counters in ways that more resemble chicken than seafood. The salmon industry, from producers to distributors to retailers, is increasingly becoming more like a food industry than a seafood industry. A very important structural change in this process is the growth of supermarket chains as the main retail outlets.

7.2.6   *Salmon and supermarket chains: a marriage made in heaven*

Beginning in the late 1980s in northern Europe and the USA, a revolution took place in the retail sector. The size of retail outlets increased as supermarkets became more prevalent, hypermarkets were introduced, and these outlets were organised into supermarket chains. There are several reasons for this, including the increased mobility of consumers that allowed retailers to exploit economies of scale in the shops, as well as scale economies in marketing, logistics and distribution. For aquaculture, and particularly for salmon since this was the only species available early on, these factors provided an important competitive advantage over wild fish. Fresh wild fish are supplied by fishermen at the mercy of seasonal cycles, are of inconsistent quality, vary in size, and are a challenge to handle in the distribution chain. In fact, in northern Europe, several supermarket chains removed their fresh fish counters around 1990. In contrast, at the end of the 1980s salmon farmers had reached production volumes that allowed them to supply salmon consistently over the year, the kind of product that was suited to the logistics systems of supermarket chains. With salmon, it was possible to plan marketing campaigns several months in advance and know that the required volume was available.

As such, the supermarkets boosted demand growth for salmon as they exploited the competitive advantage of farmed salmon. The shift in retail sales of fish and seafood has been dramatic during the last two decades.

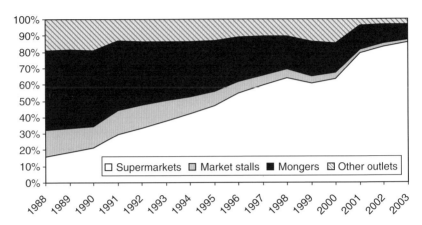

Figure 7.1   Retail shares in the UK. (*Source*: Sea Fisheries Industry Authority (SFIA))

In many European countries, the supermarket chains had more than 70% of all retail sales of fish in 2000, up from less than 20% in the late 1980s. Their share continues to increase as Figure 7.1 shows for the period 1988–2003 in the UK. In 1988, fishmongers and market stalls had 65% of salmon sales. By 2003, this was reduced to 10.7%. During the same period the retail chains' market share increased from 16 to 86%. Although up-to-date information is not available as the data series has been discontinued, according to industry sources by 2009 the chains' share might have reached 88%. This has clearly increased the demand for salmon, and also increased its competitiveness as not only production costs but also distribution costs have been reduced.

There are no reliable data that allows research into how much the retail price of salmon has been reduced due to improved logistics. However, there are some figures that give some indications. Figure 7.2 shows retail prices for salmon fillets in France for fishmongers and supermarkets. Supermarkets clearly sell at a lower price than fishmongers. On average, the difference is €3.24/kg, and the margin does increase over time. From the late 1980s to 2000, the supermarkets' share of retail sales increased from less than one-third to over 80%. It is likely that they are outcompeting the fishmongers because they can pass on at least a part of the savings from lower distribution costs to their customers in the form of lower prices.[31]

A comparison of fresh salmon and fresh cod is of interest. Cod is one of the high-volume wild species sold fresh in Europe, and as such it has a fairly efficient supply chain relative to many other wild species. Norwegian fishermen receive between 10 and 15% of the retail price in the UK. For

---

[31]   The same relationship exists with fillets, but not necessarily with more aggregated prices. This is because the increased value added in total salmon sales distorts the more aggregated prices.

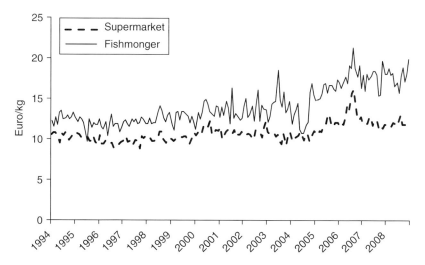

Figure 7.2   Retail price for salmon fillets in France, 1994–2008. (*Source*: Ofimer)

salmon, the producer receives almost 50% of the retail value in the UK regardless of whether a Norwegian or Scottish producer price is used. The margin in the supply chain to France produces a similar number. If cod had the same efficient logistics as farmed salmon, its price could be reduced by 70%. This could of course help cod aquaculture tremendously, but it is primarily a challenge for the competitiveness of wild fish. All the advantages for salmon are disadvantages for cod and other wild fish. The supply is seasonal and unpredictable, there is often idle capacity in the logistic systems, there is uneven turnover, and too often a substantial part of the cod taken into a shop is not sold, leading to a loss for the retailer.

7.3   The size of the market

The markets for salmon and salmon trout have been subject to extensive econometric analysis, both demand analyses and market integration analysis. The demand analyses aimed to provide quantitative estimates of the effects of changes in the own price, in the prices of substitutes, and in consumer income on the demand for salmon products. Analyses of price relationships aim to determine to what extent markets for different salmon products are integrated, i.e. whether the prices of the different products follow each other over time.[32]

---

[32]   The advantage of analysing price relationships, compared to demand analysis, is smaller data requirements, since quantity and income data are not required. The drawback is that one obtains less information than traditional demand analysis provides.

Most farmed salmon is sold fresh. Market delineation studies of salmon markets have found that there is no separate market for fresh salmon in the EU or globally (Asche & Sebulonsen 1998; Asche 2001) and accordingly it is a global market. The price level differs, with higher prices in Japan than in the EU or the USA, due to higher transportation costs, but the prices follow the same pattern over time. An integrated market for fresh salmon implies that changes in price in the EU market will also affect prices in other major markets. However, while price differences in, say, France and the UK are corrected relatively quickly as there are a number of firms shipping salmon to both markets, it takes more time between different continents. In particular, it seems that the link between the North American market for fresh salmon, primarily supplied by Canada and Chile, and the markets that are primarily supplied from Europe is not so strong. However, even if there can be substantial independent price movements in the short run, prices will become aligned in the long run.

Market integration studies also indicate that farmed salmon competes closely with wild. In fact, the different species seem to be such close substitutes that the Law of One Price holds (Asche et al. 1999b, 2005). As most farmed product is sold fresh, while most wild is sold frozen, this also implies that there is a close relationship between fresh and frozen. Further, salmon trout clearly also belong in this market (Asche et al. 2005). However, a number of studies indicate that there is little or no substitution between salmon and other fish species (Gordon et al. 1993; Jaffry et al. 2000; Asche et al. 2002, 2004; Nielsen et al. 2007, 2009).

An important issue in studies of the demand structure for salmon is the substitutability of different forms of salmon, and substitutability with other food products, such as other fish and meat products. For example, how many percent will the demand for salmon increase if the price of other fish products (such as turbot or cod) increases? Substitutability has been analysed empirically by means of econometric demand models. Such models estimate the cross-price elasticity, which is defined as the percentage change in the demand for a product when the price of another product increases by 1%. Cross-price elasticity estimates for fresh and frozen salmon in the EU market suggest moderate to high degrees of substitutability between different product forms of salmon (Asche et al. 1997, 1998).

It is worthwhile to note that there seems to be few if any clear substitutes for salmon. Obviously, with the large increase in production, salmon must have won market share somewhere. However, there is little evidence with respect to where. Through the 1980s salmon was still regarded as a high-end product, and there is some evidence that there was substitution against other high-valued species (Bjørndal et al. 1992). As the price declined, this substitution disappeared, and no new substitutes have been found. Indeed, as there are few other products that have a similar price decline, that would be as expected. Rather, it seems as though salmon has won market share from a large number of products, and because the share in each case is very small, it cannot be measured. In fact, in some markets there is evidence that salmon has boosted demand for other seafood, indicating a complementary relationship. This is a surprising result, but nevertheless it makes sense. In northern Europe the number of fresh fish counters and fishmongers has been reduced, due in part to the substantial cost of keeping them open with expensive logistics. The stable supply and sales of salmon can in some of these markets make a difference, allowing the outlet to be kept open. In this way salmon sales also increases sales of other species.

According to most studies salmon demand is fairly elastic in own price, so that a 1% decrease in the price of salmon will lead to a more than 1% increase in salmon demand. However, the magnitude of the elasticity is becoming smaller as the supply has been increasing, and for some species or product forms it already seems to be inelastic. If the supply of salmon continues to increase, it is most likely only a matter of time before aggregate demand for salmon becomes inelastic. In fact, it may already be inelastic, since most studies use older datasets, and elasticities reported at the mean will contain substantial weight from data from the 1980s.

The effect of changes in income on the demand for salmon has also been estimated. In general the studies find that salmon demand is income elastic, so that a 1% increase in income increases demand by more than 1%. To what extent this will be the case in the future depends on whether salmon maintains its historical reputation as a luxury product. Distribution through supermarkets, and new product forms, may change consumers' perceptions of salmon and consequently their tendency to purchase salmon as income increases.

7.4 Salmon marketing

Marketing is the means by which companies try to increase market size. It comes in a number of forms such as advertising, promotion and public relations work. A challenge for the salmon industry (in common with many other food producers) is that each farm is relatively small. With limited resources, each company therefore could have little impact by itself. In all countries there are industry organisations that carry out public relations work. In some countries, there are also organisations that collect a fee to carry out marketing on behalf of the companies in the industry. Such marketing is known as generic marketing, as it emphasises general messages about the attributes of the product (e.g. milk is good for you, eat more salmon) and not the producer's brand. However, in many cases these organisations are working for the producers in a single country, and country of origin labelling is therefore often an important part of generic marketing.

When production of farmed salmon took off in the early 1980s, the industry could exploit existing distribution channels. However, in order to find buyers for rapidly increasing volumes, a major challenge for the industry was to find new distribution channels and means of advertising to make more consumers aware of the benefits of salmon. As with other fish, it has been difficult to establish strong brand names, because one supplier's investment in advertising provides positive externalities for other suppliers. This market failure led the Norwegian industry to organise generic marketing for salmon through their joint sales organisation, and similar organisations exist in other producing countries. This effort was largely terminated in 1991, when the Fish Farmers Sales Organisation went into bankruptcy (see Chapter 3). Soon thereafter, a fee on all seafood exports, including salmon, was introduced to be used mainly for generic marketing. The fee, which for salmon was 0.75%, was collected and administered by the Norwegian Seafood Export Council. In 1997, generic marketing increased substantially due to an increase in the fee to 3% of salmon exports to the EU. This was in conjunction with a trade agreement with the EU (the so-called Salmon Agreement, which is further discussed in section 7.5). The tax was earmarked for generic marketing, meaning it would also benefit the UK and Ireland. Because of this tax the Norwegian Seafood Export Council spent NOK235 million on generic marketing of Atlantic salmon in the EU and other markets in the period 1997–1999. In the following years almost NOK390 million was used annually for marketing purposes. However, this was reduced in 2003 when the Salmon Agreement expired.

The first priority of the Norwegian Seafood Export Council is to ensure continued growth, and also maintain Norway's position in the EU market. Japan and the USA have been defined as core markets with large potential, while Asia and eastern and central Europe are viewed as important emerging markets. An important trend in salmon marketing has been penetration into new market segments through the distribution of increasingly cheaper fresh and frozen products to supermarkets. This has opened up a new market of consumers who consider salmon a substitute for chicken and other meat products. In the future, increased concentration and vertical integration, increased traceability of the salmon through the value chain, and increased product differentiation (e.g. dinner-ready frozen products) should to a larger degree facilitate private marketing and establishing private labels.

7.4.1 Health benefits, food scares and environmental concerns

It is not only the information provided by producers and their organisations that will influence potential buyers' perceptions of salmon. There are also a large number of other agents who provide information that influences consumer perceptions about how desirable it is to consume salmon (or seafood), and as such influence market size both positively and negatively.

In general, seafood (and particularly fat fish like salmon) is perceived to be good for health, and the positive impact is well documented, as reviewed by Mozaffarian and Rimm (2006). Information indicating that a product is good for people makes it more likely that they will buy the product, and as such has the same impact as advertising. Seafood and fish like salmon have been shown to have positive effects on reducing negative cardiovascular and neurological outcomes in adults and on early neurodevelopment. Public agencies and doctors provide such information and articles are published in the press. This positively influences demand for seafood. However, seafood (and again fat fish like salmon) also contains contaminants such as mercury and polychlorinated biphenyls that increase the risk of cancer. When such information is provided in the press, it will reduce demand for these products. The fact that the information provided is contradictory makes it hard to assess its total impact on the demand for seafood. However, there is evidence that specific studies have a measurable impact on demand. While the information is contradictory, medical researchers seem to agree that the positive health effects outweigh the negative. For instance, Mozaffarian and Rimm (2006, p. 1985) state 'levels of PCBs and dioxins in fish are low, similar to those in several other foods, and the magnitudes of risks in adults are greatly exceeded by benefits of fish intake and should have little impact on individual decisions regarding fish consumption'. Hence, it is likely that consumption of seafood and fish will continue to be associated with positive health effects, and influence demand positively.

As discussed in Chapter 5, environmental concerns are related to aquaculture, including salmon aquaculture. While this is primarily a production concern, it also becomes a market concern if information about the production process influences consumer choices. Many consumers are concerned about the environmental sustainability and impact of the production process of the food they are buying. They will stop buying products that do not meet their standards. For every consumer who decides not to buy a product, there is a reduction in market size, and if a number of consumers make this choice, it can influence market size significantly. Conversely, products with a good environmental reputation can increase market size. Because many consumers have environmental concerns, there are a number of sources of information with respect to the environmental impact of purchases. This includes marketing and public relation campaigns targeting specific products, as well as retail chains that claim to only source sustainable products, as for instance Whole Foods in the USA. For wild fish, the Marine Stewardship Council is providing an ecolabel to certify that fish have been harvested in a sustainable manner. While this does not directly influence aquaculture, it can have an impact to the extent that consumers perceive wild and farmed fish to be competing. There is evidence that they are, although the direction is not clear.[33] Anderson and Bettancourt (1993)

---

[33] To prevent this, the organisation Global Aquaculture Alliance has developed a best standard certification scheme for aquaculture products. The organisation Friends of the Sea issue ecolabels for both wild and farmed seafood.

indicated that US consumers at the time preferred farmed fish to wild, presumably because of better quality control. However, as more concerns have been raised about the impacts of aquaculture, there is a perception that consumers in general now prefer wild to farmed fish. There is a surprising lack of interest in this question in academic studies, although there is a large grey literature (a review can be found in Knapp *et al.* 2007).

Food scares and food safety will have similar impacts. A positive safety perception will lead to increased demand. When a food scare relating to a given product hits the news, it will reduce demand. Scares for competing products can increase demand as consumers shift demand away from the problematic product. However, it should be noted that the impact of food scares does not tend to be prolonged, and in this respect consumers seem to have short memories.

7.5 Trade restrictions

As discussed in Chapter 3, a number of regulations have been implemented in salmon-producing countries, with a number of different objectives. A feature of the salmon market is that some producers are located within two of the main consuming areas and operate under trade agreements, namely Canada and the USA in NAFTA, and Scotland and Ireland in the EU. This has led to a number of trade conflicts where, depending on the point of view, NAFTA and EU producers either receive just protection from unfair competition or receive government support from protectionist measures to gain a competitive advantage. This adds a special element to the market structure as in certain periods some producers have experienced regulated market access.

Before discussing trade conflicts in relation to the salmon industry, it is useful to give a brief review of the measures that a country (or trading block) can take to protect domestic producers. The General Agreement on Tariffs and Trade (GATT) and the World Trade Organisation (WTO) have established substantial limitations on what trade-restricting measures a member country can implement, as the primary motivation for these agreements is to reduce trade barriers. The agreements prevent a country from arbitrarily introducing any measure considered to be trade restricting.

The motivation for tariff-reducing agreements is that increased trade is welfare increasing. However, producers in importing countries will lose with increased trade. It is therefore in their interest that the domestic market is protected, but the GATT and WTO agreements in general prevent this, as member countries are not allowed to increase trade barriers. This is to ensure that welfare-enhancing competition is not hindered by trade barriers. However, there are exceptions that allow countries to erect trade barriers in response to what are labelled as unfair trading practices. These come in three main categories: dumping, subsidies, and market disturbances. They can be countered with anti-dumping duties, counter-vailing tariffs and safeguard measures.

Dumping involves producers selling their product in a market without covering costs. This can be a profitable market strategy if it bankrupts the domestic industry, and the dumper can subsequently take over the market and increase prices when the domestic firms have disappeared. As such practices are regarded as unfair (and are prohibited for domestic firms in anti-trust laws), a country is allowed to protect its own industry by impos-ing an anti-dumping duty to prevent these practices.

The effects of subsidies are in many ways similar to the effects of dump-ing, except that it is the exporting firms' government that allows the firms to sell at an unfair price because the subsidy gives them lower costs. There are many reasons for governments to subsidise an industry, but subsidies are considered unfair if they achieve competitiveness at the expense of an industry in another country.

Based on these definitions it should be straightforward to determine whether a company dumps or receives subsidies. However, this is often not the case. Even with access to a firm's accounts it is often difficult, if not impossible, to determine production costs. For instance, it is known that the price of salmon tends to increase with weight. Is the low price to a spe-cific country due to dumping or because small fish are being shipped? Similarly, is the building of public infrastructure a subsidy? It is if for instance a road is built to reduce transportation costs for a salmon farm, but not if it is built to provide better communications for a small community where there happens to be a salmon farm. Moreover, since the information available to the agency determining whether dumping or subsidies have taken place varies, there are different ways to compute dumping margins and whether subsidies are provided. This also leaves room for substantial discretion for the agency with respect to what margins it finds, and some commentators claim that it is primarily a policy exercise, since it can often find whatever it wishes.

Safeguard measures are somewhat different, in that they address an unfair outcome (caused by a market disturbance) rather than a cause. A safeguard measure can be implemented when imports of a product increase so rapidly that they threaten the stability of the market. Safeguard measures can be implemented when imports increase by more than 10% during a period. The underlying motivation for safeguard measures is an understanding that increased competition should happen gradually, and when sudden market shocks occur, there must be reason for concern.

If the exporters in a country are found guilty of dumping, receiving sub-sidies or increasing exports so fast that market stability is threatened, the importing country has a substantial menu of measures that can be taken to prevent or reduce the effects of these problems. The tools come in three main forms: tariffs and duties, import quotas and minimum import prices. Each type of measure can be implemented in a variety of forms, and combinations can also be used. However, while measures can be imple-mented if a trade problem is discovered, they may not be. The agency that investigates these matters must determine whether the wrongdoing causes substantial harm before any measures are implemented. In this exercise, the benefit to consumers is weighed against the loss to producers.

Of the three measures discussed, anti-dumping is the most commonly used. While often sensible and necessary, and usually straightforward to determine when a case of dumping has taken place, anti-dumping can also be misused. Many authors claim this is the case, as anti-dumping was rarely used until the late 1980s.[34] The number of anti-dumping cases exploded when trade barriers were reduced because of GATT, and tradi-tional trade-restricting measures were no longer available (Prusa 1996). In more than 90% of complaints, some kind of measures are implemented, either directly by the plaintiff's government or in the form of so-called vol-untary trade measures that the 'accused' agree to, since they will most likely be found guilty anyway. Hence, it is far from clear whether dumping is used primarily to correct unfair trading practices or is used as a protec-tionist measure.

In addition to the explicit trade measures that can be taken under the WTO agreement, there is also another category of measures that are not strictly trade measures but which often have the same effect. These are measures that are taken to uphold, for example, food safety standards in a country, or to maintain control of imports. For instance, restricting inspec-tion of fresh fish at the border to normal working hours may sound innocu-ous, but if inspections have previously been available around the clock this can significantly limit trade. Similarly, food safety is a significant concern in many countries, and at times imports from specific countries are restricted or prohibited for safety reasons. While these concerns clearly can be legiti-mate, there is an impression that they are often implemented for other rea-sons. Non-tariff trade barriers then take a role similar to anti-dumping, even though they represent legitimate and necessary concerns for a country under given circumstances. However, as with anti-dumping, the use of such measures increased significantly after the GATT limited the use of tariffs to restrict trade, and critics (and exporting countries) will often claim that they are misused. In the salmon market, Norway has had several issues with the EU with respect to veterinary control at the border, and there was a period in 2006 when exports to Russia were suspended due to Russian food safety concerns. As can be seen in Table 6.10, this clearly impacted Russian imports of fresh salmon. While the importing countries claimed that these measures were legitimate, Norwegian exporters certainly thought otherwise. And although the salmon market seems to have had more than its share of anti-dumping complaints, there have been fewer cases where non-tariff trade barriers have been a concern, relative to many other species.

---

[34]   See the article 'Unfair Protection', *The Economist*, 7 November 1998.

7.5.1 *Trade restrictions in the salmon market*

Rapid increases in salmon production, and variations in profitability as production costs and prices have decreased, have opened the salmon industry to trade conflicts. There have been a number of trade disputes during the last decade, in both the USA and the EU. Attempts to protect domestic producers from the full consequences of tough competition have been perceived to be unfair by other producers. Norwegian producers have been the primary targets because of their large share of production. However, Chile has also been increasingly targeted. An overview is provided in Table 7.1.

In the EU, complaints comprise two varieties: informal complaints in cooperation with Scottish and Irish politicians to the EU Commission, often based on safeguard rules; and formal dumping complaints. So far, the first strategy has been the most successful. The EU Commission has on several occasions implemented minimum import prices for all Atlantic salmon, although Norwegian producers are the primary targets. Scottish and Irish farmers have also argued in favour of import quotas, but the Commission has not been willing to impose this measure with the exception of the safeguard measures in the autumn of 2004. However, after pressure from the EU, and to avoid more serious measures, the Norwegian government, and since 1997 the producers themselves, have implemented several voluntary trade restraints. These include feeding stops and a system with feed quotas in order to restrict production. These have also been more formalised in the so-called Salmon Agreement between the EU and Norway in the period 1997–2003. Following termination of the agreement in 2003 there has also been a string of safeguard and other measures. However, after a WTO panel in 2007 published its preliminary findings, all EU measures restricting Norwegian salmon were terminated. In total, there have been measures that regulate imports to the EU for more than three-quarters of the time since 1990. However, although the measures surely have been costly at times, the tremendous growth in imports and consumption in the EU indicates that the total effect has been limited.

Irish and Scottish farmers have on several occasions, in 1989, 1991, 1996 and 2004, made formal dumping complaints against Norwegian farmers.[35] The first three complaints were not successful. However, the complaint in 2004 led to safeguard measures, and as these were challenged before the Commission, to a short period of anti-dumping duties in 2005, prior to new safeguard measures again being implemented. Norway filed a complaint about the measures implemented by the Commission in 2005 to the WTO. The WTO ruling did not give either party clear support, but it led to all EU measures against Norwegian salmon being terminated in 2007.

---

[35]   Please note that these dates also correspond well with periods with poor profitability in the industry, as indicated in Figures 4.1 and 4.5.

Table 7.1   Trade conflicts in relation to salmon.

Year	Action
1989	December: Scottish farmers send a formal accusation of dumping to the EU
1990	March: the US International Trade Commission opens an investigation of Norwegian salmon following a petition from the Coalition for Fair Atlantic Salmon Trade
	June: USA introduces a temporary subsidy fee of 2.96% on all imports of Norwegian salmon
	October: USA introduces a 'penalty' duty of 2.96% on all imports of Norwegian salmon
	European Commission suggests a 'penalty' duty of 11.32% on Norwegian salmon
1991	February: final subsidy duty of 2.27% on imports of Norwegian salmon to USA, and final 'penalty' duty on imports of Norwegian salmon to USA, 23.8% on average
	March: EU renounces the introduction of the 'penalty' duty of 11.32% on Norwegian salmon due to the freezing programme
	November: EU introduces minimum price on imported Norwegian salmon until beginning of March 1992
	December: Scottish farmers send their second accusation of dumping to the EU
1992	January: 'Stop feeding' action, organised by Norwegian Fish Farmers Association (NFF)
1993	October: 'Stop feeding' action, organised by NFF
	November: EU introduces minimum prices until 31 January 1994
1994	September: Scottish farmers withdraw the dumping accusation against Norway sent at the turn of the year 1991/1992. The letter was sent just before the Norwegian EU vote
1995	March/April: feeding stop, organised by NFF
	December: EU introduces minimum price on Norwegian salmon, under surveillance until 30 June 1996
1996	August: EU opens anti-dumping lawsuit against Norway
1997	March: EU Commission suggests a 'penalty' duty of 9.88% and a subsidy duty of 13.7% as a result of the dumping and subsidy investigation
	May/June: the Salmon Agreement is adopted by the Commission on 1 July 1997 with a minimum price of €3.25
	June: Coalition for Fair Atlantic Salmon Trade in the USA files an anti-dumping complaint against Chile
	July: temporary duties on imports of Chilean salmon to the USA
1998	January: final duty on Chilean salmon at 5.19% for smaller producers, and between 2.24 and 10.91% for larger producers
2002	March: European Commission suggests a replacement of the Salmon Agreement with a duty. After negotiations, the agreement is extended. Anti-dumping lawsuits are opened against Chile and the Faeroe Islands
	December: the Commission suggests termination of the Salmon Agreement. No action is taken against Chile and the Faeroe Islands
	December: anti-dumping case filed against Norway and the Faeroe Islands
2003	May: the Salmon Agreement is terminated
	June: the duty on Chilean salmon to the USA is abolished
	September: temporary 'penalty' duty of 21.4% on Norwegian trout
2004	February: request from Ireland and the UK on temporary safeguard measures on Norwegian, Faeroese and Icelandic salmon

Table 7.1 *(continued)*

Year	Action
	March: permanent 'penalty' duty of 19.9% on Norwegian trout
	May: safeguard committee rejects the proposal from the Commission of a 13% 'penalty' duty. The Commission withdraws the suggestion
	August: EU Commission introduces temporary safeguard measures in the form of an import quota and duty on the amount exceeding the quota
	October: anti-dumping lawsuit is filed against Norway from EU farmers
2005	January: new temporary safeguard measures
	May: Norwegian salmon exporters are found guilty of dumping. Temporary anti-dumping measures are introduced from June
	June: temporary anti-dumping measures (minimum import price) are introduced, and the tariffs introduced in May is revoked
2006	January: temporary measures from June 2005 are made permanent for a 4-year period
	Norway files a complaint to the WTO
2007	July: all EU measures against Norwegian salmon are terminated after preliminary findings from the WTO panel published

Following a dumping complaint from salmon farmers in the USA in 1989, of which Norwegian farmers were found guilty, imports of fresh salmon from Norway have faced a countervailing tariff that is on average 26.7%, with some variations depending on the firm.[36] The introduction of the tariff in April 1991 effectively closed the US fresh salmon market for Norwegian producers. However, the exclusion of Norwegian fresh salmon seems to have had little benefit for US domestic farmers, as Chilean and Canadian farmers appear to have taken over most of the Norwegian market share.

Farmers in the USA filed a new dumping complaint against Chile in 1997 and in June 1998. The US Department of Commerce found Chilean farmers guilty of dumping salmon on the US market. After having faced a temporary duty since January 1998, Chilean salmon faced an anti-dumping duty of 5.19% for most firms from June 1998. The duty was abolished in 2003, and again the effect on US farmers was negligible. The US measures certainly had significant short-run costs, but in total the volume of imports and increased consumption indicate that, as in Europe, they did not limit market development to a substantial degree.

---

[36] The duties are an equalisation tariff of 2.27% and a countervailing duty varying from 15.65 to 31.81%. Recently, a few Norwegian firms have been able to prove in US courts that they did not dump and have been exempted from the duties.

Appendix: a market model

Market interactions are the core of microeconomic theory. A formal, albeit simple model of market interactions will be developed to shed light on some of the relationships under investigation. In particular, an interesting relationship is that an important subset of the potentially competing goods to farmed fish, captured fish, have a backward bending supply schedule (Anderson 1985).

Consider a very simple model with two goods, the aquaculture (A) product and a potentially competing product (O). Let q_i^D be the quantity demanded, p_i and p_j the prices of the two products, and I consumer expenditure or income. The demand for the two products can be written as:

$$q_i^D = a_i - b_i p_i + c_j p_j + d_i I \tag{1}$$

where $i = A$ and $j = O$ if it is the demand for the aquaculture product, and $i = O$ and $j = A$ if it is the demand for the competing product. For the interaction between farmed and other products the parameter c_j is of key interest, as this gives the strength of the substitution effect. In particular, if $c_j = 0$, there is no substitution effect and therefore no market interaction. If $c_j > 0$, the two goods compete and are substitutes, whereas if $c_j < 0$, the goods are complements.

The demand for a product is often shown graphically in price–quantity space as a demand schedule (Figure A1). The demand is downward sloping since the buyers will purchase more, the lower the price; also, as price is reduced, new consumers will enter the market. The demand schedule also gives the size of the market for a product, as the combination of price and quantity gives the total value of the market. Market size changes when the demand schedule shifts. In the figure a positive shift is shown (curve D'). This can be caused by increases in the prices of competing products and increased income. Finally, it can be caused by a number of non-economic factors that are lumped together in the constant term a. Marketing, for instance, will normally work as a positive shift in the demand schedule, and will in this model show up as an increased constant term. However, it is worthwhile to note that marketing can also influence other parameters in the model so that the schedule rotates.

Similarly, when assuming one input factor, the supply of farmed fish will be a function of its price and the price of the input factor. Let the supply of farmed fish be denoted as q_i^S, the output price as p_i, and the price of the input as w_i. This can be written as:

$$q_i^S = m_i + n_i p_i - o_i w_i \tag{2}$$

The main cause for the increased supply of farmed fish is productivity growth. This may take two forms: productivity growth in the farming

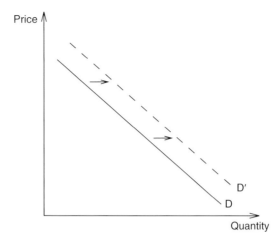

Figure A1   Demand.

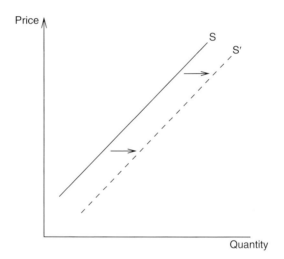

Figure A2   Supply.

operation and productivity growth for the suppliers of input factors. The first corresponds to a reduction of m_i, while the second results in reduced input price(s), w_i. Productivity growth leads to a downward shift in the supply schedule.

The supply schedule can also be shown graphically (Figure A2). The supply schedule is upward sloping, as producers will be willing to produce more, the higher the price (p_i). The supply schedule will shift due to changes in economic factors such as input factor prices as well as productivity growth. Productivity growth is normally captured by the constant term m.

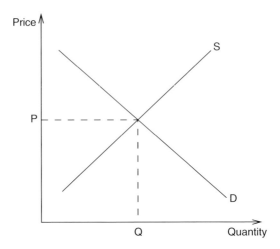

Figure A3   Market equilibrium.

If this is reduced, the supply schedule shifts downward, and the suppliers will be willing to supply more of the product at any price. Technological progress can also result in changed parameters on the output price or the input prices, so that the schedule rotates.

Equilibrium is where the supply and demand schedules cross, as quantity demanded at that price is equal to quantity supplied (Figure A3). If the demand schedule shifts out, the market size will increase. The market equilibrium will also shift as it will have to be on the new demand schedule. The shift will partly lead to an increase in price and partly to an increase in quantity. Exactly how much the price and quantity increase is determined by the slope of the supply schedule. Similarly, a downward shift in the supply schedule due to productivity growth will lead to a shift in the equilibrium down along the demand schedule. Hence it will lead to an increased quantity and a lower price. The fact that prices decline as quantity increases for farmed species is then an indication that productivity growth is faster than market growth.

Possible market interactions are illustrated in Figure A4, where demand and supply both for farmed fish and for a potentially competing product are shown. Assume that the prices are normalised so that they initially are equal for the two products, and that the supply of farmed fish shifts downward due to productivity growth. This leads the supply of farmed fish to increase from q1 to q1' and the price to decrease from p1 to p1'. The effect on the market for the potentially competing product depends on the parameter c_A in the demand equation for this good, since this parameter determines the cross-price effect of the competing product with respect to a price change for farmed fish. If this parameter is zero, there will be no effect: price and quantity demanded remain at p2, q2, and there is no market interaction. If the parameter is positive, implying that the two products are

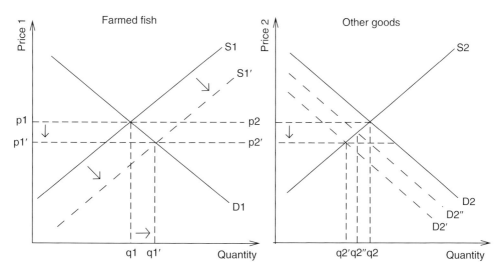

Figure A4   Potential market interaction between farmed fish and traditional goods.

substitutes, a price reduction for farmed fish will lead to a downward shift in the demand for the competing product. At most, the demand schedule for the competing product can shift down sufficiently for the relative prices to remain constant. This corresponds to the demand schedule D2′ in Figure A4, with quantity demanded reduced to q2′ and price to p2′. If there is a weaker substitution effect, the demand schedule for this product will shift down to D2″ for example. This gives a reduction in price and quantity demanded, but the shift in demand is not sufficient to keep the relative price stable.

If the competing product is a wild-caught fish species, the supply will also be a function of the biological characteristics of the stock and/or the regulatory system. In open access equilibrium the supply schedule is backward bending. Using the same bioeconomic model as Anderson (1985), the supply of wild-caught fish can then be written as:

$$q_i^S = \frac{rC}{p_i}\left(1 - \frac{C}{p_i K}\right) \qquad (3)$$

where r is the intrinsic growth rate of the fish stock, C is cost per unit of effort in the fishery and K is the environmental carrying capacity for this fish stock. If the fishery is managed with a quota, and based on biological considerations only, this makes the supply schedule vertical.

Market effects when the competing product is wild fish are illustrated in Figure A5. Figure A5 differs from Figure A4 in that the supply schedule for the competing product, captured fish, is backward bending. If the fishery is on the upward-bending part of the supply schedule, the description of the effects is as for a conventional product. The most interesting situation is if the fishery is on the backward-bending part of the supply schedule. This

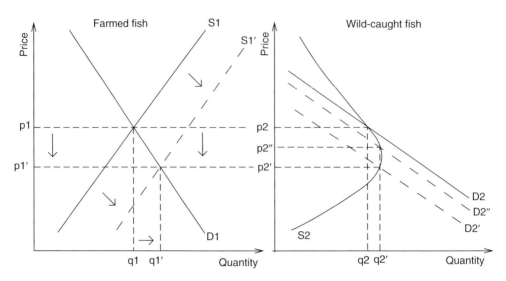

Figure A5   Potential market interaction between farmed fish and wild-caught fish.

may also be representative for many fisheries, as a large number of wild fish stocks are fully exploited or over-exploited (FAO 2009).

Assume again that productivity growth gives a downward shift in the supply schedule for farmed fish, reducing price to p1' and increasing quantity supplied to q1'. The effect on the market for the potentially competing product again depends on the parameter c_A. As above, if this parameter is zero, there will be no effect. If the parameter is positive, implying substitutability, a price reduction for farmed fish will lead to a downward shift in the demand for the competing product. At most, the demand schedule for the competing product can shift down sufficiently for the relative prices to remain constant. This corresponds to the demand schedule D2', which implies that the quantity demanded of this good is increased to q2' while the price is reduced to p2'. If there is a weaker substitution effect, the demand schedule for this product will shift down to D2" for example. This gives a reduction in price and an increase in quantity, but the shift in demand is not sufficient to keep the relative prices stable. The main difference from the case with a conventional product is the quantity effect for wild-caught fish. When the price is reduced, this leads to lower fishing effort, which gives a higher fish stock and higher landings.

While one of the primary concerns with respect to market structure for farmed fish is often the effect of farmed fish on other products, it may also be worthwhile to comment on the effect of the market structure for possible growth of aquaculture production. Let us look at conventional products first. If production of farmed fish is increasing relative to the other product, this implies that the productivity for farmed fish production increases faster than for the other product. If the two products are close substitutes, farmed fish can then win market shares from the other product. However, if the goods are not substitutes there are no market effects, and the increase in the supply of the farmed fish will only lead to a move down the demand schedule for farmed fish. Hence, for the producers of farmed fish it is easier to expand when the farmed fish has substitutes with established markets. As observed by Anderson (1985), this situation changes if the potential substitute is fish from a fishery located on the backward-bending part of the supply schedule. The increased supply of farmed fish will then also lead to a higher supply of wild-caught fish, leading to keener competition. Anderson shows that in some cases, this keener competition may drive the farmed fish producers out of business.

No mention has been made of the possibility that farmed fish and other products may be complementary. This is mostly because it increases the complexity of the model, as there must always be at least one substitute if there is any market interaction on the demand side. However, even though complementarity is a possible effect, it is most likely not very important.

Testing for market interactions

In general, microeconomic theory assumes that there exists a marketplace constituted by a group of commodities. These commodities compete in the same market because goods are substitutable for the consumer. Whether goods are substitutes or not can be measured by estimating demand equations to test whether there are cross-price effects. If there are, the goods compete in the same market. If there are not, they do not. The most common measure of a cross-price effect is the cross-price elasticity. The most usual research approach is demand analysis, where demand equations are estimated either individually or in a system of demand equations. These studies of the demand structure focus on the price sensitivity of demand, on the degree of substitution between potentially competing products and on income/expenditure effects.

A common problem is obtaining the necessary data to estimate demand equations. Often price data are available but not quantity. Moreover, while one can often find a price that is a good proxy for market price, it is hard to obtain reliable estimates of demand equations if data are not available for the full quantity consumed in a market at different price levels. This has also led to markets being defined based on price information only. When looking at Figure A4, the intuition behind this kind of definition can be seen by looking at the effect of shifts in supply and demand schedules from the price differential. When the supply curve for farmed fish shifts, the price changes, and this can have an effect on the price of the other good. If there is no substitution effect, the demand schedule does not shift and there is no movement in price. If there is a substitution effect, the demand schedule shifts, and the price shifts in the same direction as the price of farmed fish. At most, the price of the other product shifts by the same percentage as the price of farmed fish, making the relative prices constant so that the Law of One Price holds.

Hence there are at least two ways of testing for market integration, or if two or more products are substitutes. One is to estimate the demand function for a product and to test for cross-price effects. Alternatively, one can look at the effects only in the price differential, where one can test whether there is a price effect (i.e. substitution) and whether the relative price is constant, i.e. whether the Law of One Price holds.

Bibliography

Since the market is at the core of economic theory, there are many good texts on market analysis and market power. An introductory text in microeconomics is a starting point, with Varian (1992) as a good example. Deaton and Muellbauer (1980) is an excellent text on demand analysis, and Martin (1993) provides a good text on industrial organisation. Many texts in international trade provide analyses of trade measures, and Prusa (1996) is a fine place to start with respect to anti-dumping.

A number of demand analyses for different segments of the salmon market have been conducted. The first studies were carried out in Canada and the USA, with focus on wild Pacific salmon and the potential competition from salmon aquaculture (DeVoretz 1982; Kabir & Ridler 1984; Anderson & Wilen 1986; Bird 1986). Herrmann and Lin (1988), DeVoretz and Salvanes (1993), Herrmann et al. (1992, 1993), Bjørndal et al. (1994), Asche (1996), Asche et al. (1997, 1998) and Xie et al. (2008) all focus on substitution between different forms of species, product forms and producers of salmon. Wessells and Wilen (1993a, 1994), Asche et al. (1997) and Eales et al. (1997) provide information on the relationship between salmon and other seafood, and Salvanes and DeVoretz (1997), Johnson et al. (1998) and Eales and Wessells (1999) investigate substitution between different salmon products and meat. Wessells and Wilen (1993b) investigate the impact of the frozen inventories on the Japanese market.

There have also been a number of market integration studies. Gordon et al. (1993), Jaffry et al. (2000), Asche et al. (2002, 2004) and Nielsen et al. (2007, 2009) investigate the relationship between salmon and other species. Asche and Sebulonsen (1998), Asche et al. (1999b, 2001a, 2005), Clayton and Gordon (1999), Asche (2001) and Asche and Tveterås (2008) analyse the relationship between different geographical markets for salmon and different product forms. Guillotreau et al. (2005) and Asche et al. (2007b,c) investigate the relationship between different levels in the supply chain for salmon.

While there are a number of studies that provide information with respect to substitutability and market size for salmon, fewer studies directly investigate the issue of market power. Jaffry et al. (2003) and Fofana and Jaffry (2008) find little evidence of market power, while Steen and Salvanes (1999) find some evidence that the Norwegian industry could exploit market power in the short but not long run as long as a mandatory sales organisation could coordinate the farmers. Asche et al. (2007c), Asche and Tveterås (2008) and Larsen and

Kinnucan (2009) look at the effect of exchange rate changes on price relationships and margins.

Bjørndal *et al.* (1992) provide the first study that tests whether generic marketing of salmon works, and confirms this. Other studies of generic marketing of salmon are in most cases closely linked to the trade issues between the EU and Norway; a series of papers is provided by Kinnucan and Myrland (2000, 2001, 2002a, 2006). There are also a number of studies of the effect of trade restrictions, generic marketing, income growth and exchange rate changes using equilibrium displacement models based on elasticity estimates from other studies (Kinnucan & Myrland, 2002b, 2003, 2005, 2007; Aarset *et al.* 2006). Anderson (1992) and Asche (2001) investigate the US case against Norway, and Virtanen *et al.* (2005) provide a study of the impact of EU's salmon measures for Finnish salmon trout producers.

The health effect of salmon and other seafood is addressed in Willett (2005) and Mozaffarian and Rimm (2006). A number of studies have investigated the importance of different types of information relative to food safety and ecolabelling, including Anderson and Bettancourt (1993), Holland and Wessells (1998), Wessells *et al.* (1999), Johnston *et al.* (2001), Wessells (2002), Roheim (2003) and Johnston and Roheim (2006). Vukina and Anderson (1993, 1994) and Guttormsen (1999) provide forecasting models for salmon prices. Finally, Wessells and Anderson (1992), Kinnucan and Wessells (1997), Kinnucan *et al.* (2003) and Asche *et al.* (2007d) provide reviews of seafood markets which compares the salmon market to other seafood markets.

Anderson (1985) provides a theoretical discussion of how aquaculture influences a wild fishery. Copes (1972) derives the backward-bending supply schedule in open access fisheries, with Bjørndal and Nøstbakken (2003) providing an empirical application.

8 Lessons for Other Farmed Species

As shown in Chapter 3, since the early 1980s the international salmon aquaculture industry has experienced growth rates that have been surpassed by few other production sectors. Global annual industry output growth has averaged 20.5% over the period from 1980 to 2008, reaching a total production of about 1.93 million tonnes in 2008. Accordingly, when measured by growth, salmon aquaculture is clearly a success story together with other farmed species like shrimp and tilapia. However, the industry has also experienced a number of challenges during this period.

Most new farmed species will reach a critical stage similar to the situation that salmon was in around 1980, when the production technology was just being mastered. Hence there should be important lessons to be drawn from salmon aquaculture. Salmon is the farmed species where nearly all economic and market research has been carried out and is therefore where the most knowledge resides. Thus it is of interest to see how the development of other aquaculture species corresponds to that of salmon. In this chapter, we first discuss the development of farmed shrimp, sea bass, sea bream, turbot and tilapia production before analysing some of the main lessons that can be learned.

8.1 Other farmed species

8.1.1 Shrimp

Shrimp is the most valuable farmed species and the seafood species with the highest trade value. Several species are farmed, with white shrimp and black tiger prawns being the most important.[37] Aquaculture production started increasing in the 1970s, and in 2008 was about 3.39 million tonnes. Production growth has, with the exception of the last few years, been slower than for

---

[37] The distinction between a shrimp and a prawn is anything but clear and also differs between markets. Here, different species of shrimp and prawns are treated as one, shrimp.

The Economics of Salmon Aquaculture, Second Edition. Frank Asche and Trond Bjørndal.
© 2011 Frank Asche and Trond Bjørndal. Published 2011 by Blackwell Publishing Ltd.

salmon. There have also been more challenges. In particular, there have been several serious diseases, the most significant being white spot disease. There have also been substantial environmental challenges. These challenges have led to substantial shifts in what have been the most important producer countries (Anderson 2003), and have also led to periods of slower growth as production has been substantially reduced in some regions.

There are several reasons why shrimp aquaculture has faced these challenges. Production technology in shrimp farming is much more diversified than for salmon. It ranges from relatively simple operations where wild-caught larvae are placed in crudely dug ponds to sophisticated closed production systems comparable to salmon aquaculture. Moreover, production in many cases takes place on land with little economic value (such as mangroves) and poorly defined property rights in countries with no environmental regulations. This has led to environmentally unsound practices in several cases. With high stocking densities, there is high exposure to disease. Moreover, in countries with poor environmental regulations, it may be more profitable to produce with high density for a few years and leave a devastated piece of land than to operate on a sustainable basis. However, technology is steadily improving and best practice producers with closed production systems operate with little risk of disease and in an environmentally sound manner. These sounder practices will most likely win market share, as the scope for technological innovation and productivity growth is highest under such conditions. There are also efforts to promote them in the marketplace. For instance, the Global Aquaculture Alliance (GAA) is providing certification of good shrimp farming practices. However, with the different technologies available, unsustainable operations are likely to remain a part of the picture, as shrimp production can be very profitable in poor countries with little or no environmental regulations.

Most shrimp aquaculture takes place in Southeast Asia, a region that has almost 90% of production. There is also substantial production in Latin America, where Ecuador is the largest producer, followed by Mexico and Brazil. In 2008, China was the world's largest producer with 1 268 074 tonnes, followed by Thailand (507 500 tonnes), Indonesia (408 346 tonnes), Vietnam (381 300 tonnes), Ecuador (150 000 tonnes) and Mexico (130 200 tonnes). Thailand is the largest exporter. As is the case with salmon, the shrimp trade is truly global. One significant difference, however, is that shrimp is primarily exported frozen. Another important difference is that harvests of wild shrimp are larger than for salmon. Shrimp landings have been increasing during the last decades, reaching about 3.5 million tonnes in 2008. In recent years, production of farmed shrimp has approached the level of wild-caught warm-water shrimp, but there are additional harvests of about 500 000 tonnes of cold-water shrimp in the North Atlantic and Pacific.

Farmed shrimp production for 1980–2008, along with the price to producers in real US dollars per kilogram, is shown in Figure 8.1. The average price in 2008 of US$4.2/kg is less than half the price in 1984. Prices stabilised in the mid 1990s when production growth was flat, then declined

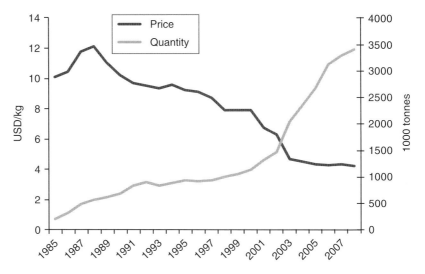

Figure 8.1   Global shrimp production and real price (2008 = 1), 1980–2008.

rapidly after 2000 while production essentially trebled over the next 5 years. As with salmon, increased production has led to reduced prices. However, the price reduction for shrimp seems to be less than for salmon, up to quite significant volumes. One likely explanation for this is that the supply of wild shrimp is relatively higher, and therefore there was a large market to win market share from. Shrimp producers have also done well in expanding market size through new sales channels.

8.1.2   Sea bass and sea bream

Sea bass and sea bream are two different species, but are produced, marketed and also consumed in a similar manner. They are often lumped together even though they have quite different appearances. There are some wild landings (annually about 20 000 tonnes combined), but these are becoming small relative to farmed production. Farmed production started to grow in the late 1980s, and passed 1000 tonnes for each species in 1987. In 2008 about 133 000 tonnes of sea bream and 116 000 tonnes of sea bass were produced.

Sea bass and sea bream are Mediterranean species, and most farming operations are located there. Although farming was pioneered in Italy and France, Greece has taken over as the largest producer reaching close to half the production by the late 1990s and over half by 2003.[38] However, despite a continued increase in Greek production, the country's share of production is declining due to a rapid production increase in Turkey. Most of the production has traditionally been located in Europe and within the EU. However, in recent years production has expanded rapidly

---

[38]   Of a total sea bass production of 116 000 tonnes in 2008, Turkey produced 49 270 tonnes and Greece 35 036 tonnes. Of a total production of sea bream of 133 026 tonnes in 2008, Greece produced 51 957 tonnes, Turkey 31 670 tonnes, Spain 22 286 tonnes and Italy 8400 tonnes.

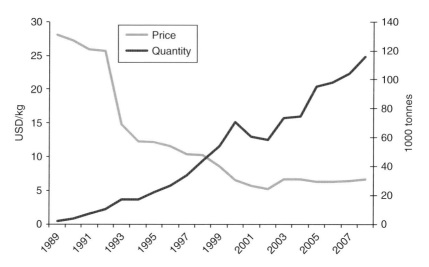

Figure 8.2 European production of farmed sea bass and real price (2008 = 1), 1989–2008.

in non-European countries. Turkey is most important, with a total quantity similar to that of Greece but with a larger production of sea bass in 2008. Farming of these species has started and is increasing in other non-European Mediterranean countries. Italy is the largest market for sea bass and takes about half the production, while Italy and Spain each consume about one-quarter of sea bream production.

The production of sea bass and sea bream can to some degree be described as salmon technology adapted to Mediterranean species, as they are the only other species produced in large-scale operations using sea pens in Europe. However, there are also important differences. Among the most important is that sea bass and sea bream are marketed as portion-sized fish at 300–500 g. This is small relative to salmon, but has the advantage that turnover becomes much higher. Given that the most valuable of the wild sea bass and sea bream landed are substantially larger, this has also separated the markets for wild and farmed fish. In recent years aquaculture producers have started to produce larger fish, finally targeting this market.

The development in production and real prices for sea bass and sea bream are shown in Figures 8.2 and 8.3, respectively. While there are differences, it is worthwhile to note that the main feature, as with salmon, is that prices decline sharply as production increases. For sea bass there is, as with salmon, an initial period where prices were maintained despite the production increase. However, they declined rapidly when production reached about 10 000 tonnes. For sea bream, prices started declining from a production of 1000 tonnes, and took a deep hit in 1993 when the prices for sea bass also declined rapidly. The common price shock in 1993 seems to have created a relatively high degree of substitution between the species, as the main price trend and the price levels are similar since 1993.

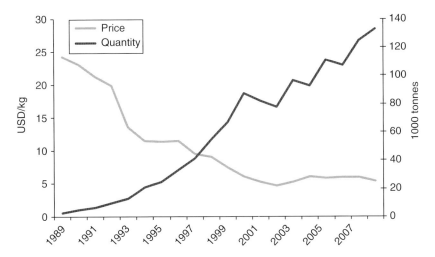

Figure 8.3 European production of farmed sea bream and real price (2008 = 1), 1989–2008.

Despite relatively limited production, prices have decreased substantially; the 2008 prices of just over \$5/kg for sea bream and \$6/kg for sea bass were in real terms only about 20–25% of their prices in 1989. Hence, the price decline has been even more dramatic than for salmon. This decline may suggest that consumer preferences are such that other fish products, or meat products, are not easily substituted for by sea bass and sea bream, and that market demand therefore represents the largest barrier to growth for these two species. They are also primarily consumed in Mediterranean Europe, and producers have not succeeded to any extent in expanding the geographical size of the market. The portion size of the fish is a limitation on the number of product forms that are marketable. Product development is very limited relative to salmon, despite the fact that sea bass and sea bream are as accessible to French processors as salmon. The fact that production in recent years (from 2002) has increased without a strong negative price effect is a sign that finally the market is being expanded. In particular, the industry has started producing larger fish, so that it is not restricted to the portion-size fish market.

The fact that production continues to grow for both species, despite declining prices, can be interpreted as evidence of substantial productivity growth, and therefore reductions in production cost.[39] However, for these species there have been periods of poor profitability, coinciding with similar poor cycles for salmon, and with a crisis in 2000–2002. The fact that Greek and Turkish producers are winning market share is an indication that they are more competitive. As a large part of Greek production is exported to other EU countries, particularly Italy, the largest market for the

[39] However, it should also be noted that in some countries production is also heavily subsidised.

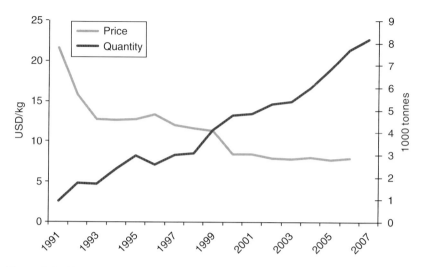

Figure 8.4 Farmed turbot production and real price (2008 = 1), 1991–2008.

species, it is likely that only Greece's membership of the EU prevents trade conflicts. As production continues to increase in Turkey and other non-EU Mediterranean countries, trade conflicts may well arise in periods with poor profitability.

8.1.3 Turbot

Turbot production was relatively moderate at just over 9500 tonnes in 2008 but it is interesting because it provides a twist to the usual development of a newly farmed species. Salmon, sea bass and sea bream are species that were very highly valued before aquaculture increased the supply, but lost that position as prices decreased. However, turbot is a species that maintains its exclusive image. Some of the explanation is of course that production has increased only moderately, as shown in Figure 8.4, from about 1000 tonnes in 1991 to 9573 tonnes in 2008. As expected, this has led to a decline in price. The price in 2008 was $8.01/kg, less than half of the 1991 price, but about two-thirds the 1993 price.

The interesting question about turbot is why production has not increased more. Many top chefs regard it as the most attractive fish species. Moreover, it has a great reputation in a much wider area than sea bass and sea bream as it is also caught (in small quantities) in northern Europe, and hence it has a larger market to tap. The reason for the limited growth in farmed quantity has to do with production technology. Turbot cannot be produced in sea pens, but must be produced in land-based tanks or raceways. This requires more investment, and makes it more difficult to increase production when market signals are positive. Moreover, the larger investment also increases capital costs, and thereby production costs, relative to species where sea

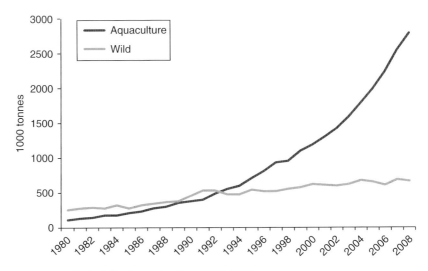

Figure 8.5   Global tilapia production, 1980–2008.

pens are used. Hence, production costs cannot decline as much as for salmon as long as this production technology is used.

The final lesson when comparing turbot with salmon is that it is not likely to remain a luxury product if production increases so much that it must compete primarily on price. On the other hand, it is virtually impossible to prevent a substantial increase in production if technological innovations reduce production costs and enhance profitability. Increased profits would be a strong incentive to expand production. Hence, for a high-valued farmed species to remain exclusive, production costs cannot be reduced too much.

8.1.4   Tilapia

Tilapia is the final species considered here. In 2008 aquaculture production reached almost 2.8 million tonnes. In addition, 668 000 tonnes of wild tilapia were landed. In terms of modern aquaculture, tilapia has been farmed for a long time. Production was over 1500 tonnes in 1950 and passed 12 000 tonnes in 1970. However, it was in the 1980s that tilapia became a major farmed species, as shown in Figure 8.5. Since the 1980s there has been a rapid increase in farmed tilapia production. Moreover, although wild landings have increased, wild tilapia's share of total supply is rapidly declining. In 2008, China was the largest producer (1.1 million tonnes) with about 40% of production, followed by Egypt, Indonesia, the Philippines, Thailand and Brazil. While better control of the production process has enabled this growth, tilapia is in many ways more interesting for what it is not than what it is, and even more for its potential.

Tilapia is an African species that is now being produced in subtropical areas globally. Production techniques differ substantially, from semi-intensive to highly intensive. While not carnivorous in nature, tilapia grows faster with fish meal-based feed. Moreover, it not only grows quickly but can also reach marketable size of 500–800 g in as little as 3 months. Tilapia's main strength is versatility. It grows well under a wide variety of conditions, and while a freshwater species, breeding has brought forward varieties that can grow in brackish water.[40] Many observers think it is only a matter of time before a variety suitable for marine aquaculture will be available.

As tilapia is a subtropical species, it is primarily produced in developing countries, which is why a wide range of production techniques are used. Relative to the quantity produced, a limited share is traded internationally. However, the share is rapidly increasing and in 2008 as much as one-third of production was traded. Only very recently has tilapia had a presence in the largest seafood markets, and that presence has been primarily in the USA. In contrast to salmon and shrimp, tilapia markets are highly segmented and diversified. Even in the USA, tilapia markets are diversified as fresh tilapia is produced locally or imported from Latin America, while frozen tilapia is imported from Southeast Asia (primarily China) at about one-quarter the price of fresh product.

A short production time gives tilapia a very high turnover, which is cost reducing. The fact that it is not carnivorous makes it likely that it will grow well on feed based primarily on non-marine ingredients. Moreover, it has only received serious research attention on a substantial scale for 15 years, and there is a huge potential for further productivity growth despite the fact that it is already a low-cost species. Finally, little has been done with respect to creating dependable and cost-efficient international distribution channels. Hence, the species has tremendous potential to become not only a globally produced but also a globally traded species. However, the pattern in quantity produced and real price is similar to what we have seen for other species (Figure 8.6). The unit price in 2008 was less than 60% of the price in 1991.

Tilapia also provides a good opportunity to look more closely at the impact of aquaculture on a developing country. Egypt, the second largest producer, is an example of a local industry where locally produced fish are consumed domestically, with no international trade. This raises several important questions. Is this an industry that in a few years will be an important supplier to Europe? Or will it disappear, as even cheaper tilapia is imported into Egypt?

Total fish production in Egypt has been increasing steadily over the past decade, primarily due to aquaculture. Currently about 50% of total supply is from aquaculture. Tilapia is the most important component of aquacul-

---

[40] Breeding is undertaken by organisations such as the Worldfish Centre and by public research institutes.

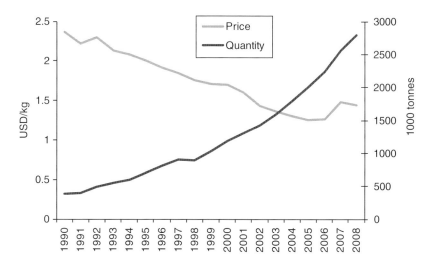

Figure 8.6   Farmed tilapia production and real price (2008 = 1), 1992–2008.

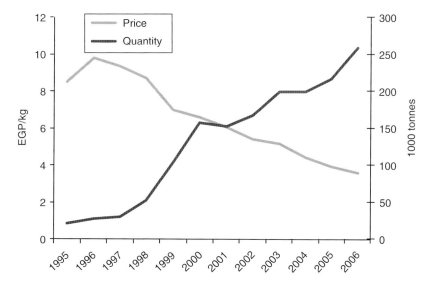

Figure 8.7   Egyptian production of tilapia and real price in Egyptian pounds (EGP)/ kg (2006 = 1).

ture and in 2008 about 386 000 tonnes were produced, or roughly 45% of Egypt's fish production. Egypt is the second largest tilapia producer in the world, up from only 21 000 tonnes in 1996. The Egyptian tilapia market is basically separated from all other tilapia markets as there are virtually no imports or exports of the species.

As expected for any market, when supply increases prices fall. This has also happened in Egypt, as shown in Figure 8.7. In the period 1996–2006 the real price fell from EGP8.53/kg to EGP3.59/kg (grade 3). Therefore, one would assume that Egyptian tilapia producers would be happy to export a part of their production if they are competitive. However this is not happening because of institutional barriers that Egypt, like many other developing countries, finds difficult to overcome together with real quality issues.

8.1.5 Other species

There are also a number of other species that are being introduced in semi-intensive and intensive aquaculture. In the last 7 or 8 years production of *Pangasius* in Vietnam has increased from about 100 000 tonnes to almost 1 million tonnes in 2008 and seems set to continue to grow at the same pace. *Kobia* is a very interesting species in Southeast Asia and the Caribbean. In northern Europe there are high hopes for cod farming. In Australia, the EU and the USA, there is a trend towards creating indoor raceways or similar systems to produce high-quality 'super fresh' fish, using species such as barramundi, kobia and tilapia.

The aquaculture industry is still so young that almost any list of important species is likely to be outdated within a few years. However, the rapid development is also an indication of the potential of the industry. Existing technologies are adapted to new species and, when adapted, give new opportunities for innovation. With the rapid pace of innovation, there are no signs that the aquaculture industry is becoming a mature industry. Innovation and the adoption of new technology are likely to continue to make farmed fish more competitive and therefore lead to increased production. Accordingly, no one is in a position to predict what is going to be the next important species, or even what species will succeed. The only thing that looks sure is that total production will continue to increase.

8.2 Lessons from other farmed species

Mastering the biological production process or, when possible, allowing nature to supply the fish is necessary for any aquaculture industry to get started. However, this is only the first step, albeit a necessary one. It is not sufficient for creating a successful food-producing industry. To be successful the species must also have suitable market conditions, and some potential for productivity improvement in the production process.

It seems to be a precondition for all new intensive aquaculture species that, initially, they should be regarded as luxury products with high prices. As prices are bound to decline when production increases, it will be greatly advantageous if price is not too much affected during the first phase of production expansion. This is not always the case for farmed species. While salmon production increased significantly before prices started to decline, it was less the case for other farmed species like sea bass and bream. It is certainly an advantage when initial prices are high as this makes it more likely that production costs will be covered. However, high prices are commonly associated with small markets, which increase the likelihood that prices will decline sharply.

Shrimp has the advantage of relatively large wild landings of the species, and hence the harvesting sector has created a large global market, with ample marketing opportunities for farmed products. Salmon has also been doing relatively well in this regard, although with substantially smaller landings of wild fish. As much of the wild salmon harvest is perceived to be of lower quality, there is a smaller market to grow in and a steeper price decline. The price reduction for sea bass and sea bream indicates that productivity growth has been at least as large as for salmon. However, production growth is still to some extent restrained, and the price decline steeper than it might have been as farmers of these species and their agents in the supply chain have been unsuccessful in expanding the market. The tilapia market in Egypt shows that even with a moderately priced species at the outset, prices can decline rapidly if market size cannot be expanded.

Shrimp, salmon and sea bass have in common initial grace periods where prices did not decline with increased supply. This can be explained by the existence of a market that was not always served by wild fish because of a seasonal landing pattern. It also indicates that by identifying underserviced markets or creating new markets, it is possible to increase production without a negative price effect.

Hence, potential producers of new aquaculture species should conduct some probing market analyses. Is there sufficient unsatisfied demand that prices will not decline initially? Is there a relatively large market so that some extra supply from aquaculture will not influence prices much, or is the market so small that even a limited quantity will make a difference? In most cases there will be a trade-off between market size and initial prices, since the market for high-priced products tends to be small. Hence, aquaculture production can be expected to influence prices more strongly when the initial price is high. However, even with moderate prices, the market can be relatively small, and therefore a moderately priced species is no guarantee against steeply declining prices. Another issue is the potential for expanding market size, either geographically or with new products, when the price starts to decline.

When assessing growth potential, markets are certainly important, but potential productivity growth is even more important. Productivity in an environmentally sustainable manner is the growth engine for any new aquaculture industry. In virtually any possible scenario, prices will start declining when a volume of a farmed species reaches the market. To remain profitable, production costs must therefore be reduced. Thus it is the combination of market size, which determines how quickly prices decline with increased production, and productivity growth, which determines how low prices can be before production becomes unprofitable, that determines how large an aquaculture industry will become. This is true independently of whether the species in question is high valued and targeted at affluent consumers in developed countries, or a low-valued species targeting self-sufficiency in a developing country.

Bibliography

There is a large literature on new aquaculture species, particularly from a biological or a development perspective. Moksness *et al.* (2003) and Leung *et al.* (2007) provide interesting examples. Certainly, biological and technological control of the production process is a necessary first step for any new aquaculture species. However, this is just part of the story. The literature from an economic perspective is more limited. Shaw and Muir (1987) and Bjørndal (1990) are early texts, but even economists did not perceive the productivity potential that control of the farming process provided. Hence, in early texts, sea ranching was perceived to be a promising technology (see Anderson 1985 and Anderson & Wilen 1985).

In the early 1990s, productivity growth was recognised as the main growth engine that enables and amplifies market growth (Kinnucan 1995; Anderson & Fong 1997; Asche 1997, 2008; Asche *et al.* 2001b). In addition to salmon, catfish in the USA was one species that was established early, and where a number of studies have been carried out (Engle 2003; Engle & Valderrama 2004). Increasingly, one can also find productivity studies for other species, such as Karagiannis and Katranidis (2000), Katranidis *et al.* (2002), Sharma and Leung (2003) and Gordon *et al.* (2008). One can also observe more interest in the markets for new species such as Young and Muir (2002), Norman-Lopéz and Asche (2008), Asche *et al.* (2009c), Norman-Lopéz (2009) and Norman-Lopéz and Bjørndal (2009).

9 Optimal Harvesting of Farmed Fish

A theoretical approach to the optimal harvesting time for farmed fish is considered in this chapter. Certain qualitative results regarding optimal harvesting are derived and analysed with regard to how they are influenced by different factors and a simple biological model, the basis for this analysis, is outlined. Thereafter, different types of costs are introduced and optimal harvesting for each case is analysed. The exposition is fairly non-technical, with an emphasis on intuitive understanding. A more comprehensive mathematical exposition is provided in an appendix that also gives examples of estimations of optimal harvesting for salmon.

The analysis is simplified in several ways. It is assumed that all plant investments are already undertaken and hence irrelevant to the decision process. Only the variable costs are relevant. Stochastic fluctuations in growth and uncertainty concerning parameter values are not considered. It is assumed that the price of fish is fixed. These assumptions may seem to be restrictive and some will be revisited later. However, in practical work one will often have to use expected values and relatively simplified models are frequently used, rather than more complex models. In Chapter 10 a practical harvesting model for fish farms will be presented, based on the models in this chapter.

The analysis is carried out in a continuous time framework. This is appropriate in a theoretical analysis such as this, as growth is a continuous process through time. This approach also constitutes the basis for harvesting models for actual fish farms. However, in such cases one would either employ a discrete time model, with the parameters updated on a regular basis, or tables based on models estimated for different scenarios. In the example in Chapter 10 a monthly harvesting model for salmon will be presented, and the connection between the two types of models will be shown.

The Economics of Salmon Aquaculture, Second Edition. Frank Asche and Trond Bjørndal.
© 2011 Frank Asche and Trond Bjørndal. Published 2011 by Blackwell Publishing Ltd.

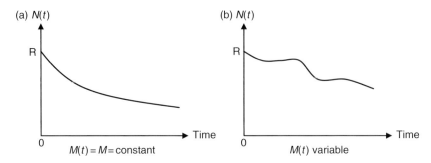

Figure 9.1   The development of the number of fish in a yearclass, N(t), over time.

9.1   A biological model

The production process in salmon aquaculture was discussed in Chapter 2. Here, the process will be described more formally. Production begins when fish are released in an enclosure such as a pen or a pond. The fish are called a **yearclass**, as all fish are of the same age. The number of fish released is for now considered given. Over time two biological processes will affect this yearclass: some individuals will grow and some will die due to natural mortality.

The development of the number of fish in a yearclass, $N(t)$, is illustrated in Figure 9.1. At time zero ($t = 0$), R fish ('recruits') are released in the farm. Over time the number of fish will decline due to natural mortality. In Figure 9.1(a) the mortality rate, $M(t)$, is constant over time, but it varies in Figure 9.1(b). There is reason to believe that mortality rates for both wild and farmed fish vary over time. For instance, the mortality rate tends to be higher at the time the fish are released in a fish farm.

Now let $w(t)$ symbolise the weight per fish at time t, e.g. measured in kilograms. As discussed in Chapter 2, the weight of the fish will grow as a function of time and other factors that will for the moment be disregarded. An example of a weight curve is given in Figure 9.2(a). In the example, fish grow towards an upper (asymptotic) value. This is always the case, as any species after a certain age will experience a decline in the growth rate and eventually death. In fish farming one can also imagine seasonal growth. An example of this is shown in Figure 9.2(b). This can be regarded as the weight functions of salmon given in Figure 2.3 extended to the full life of the fish. The time the fish reaches its maximum individual weight is given at $\tilde{t}$.

Biomass weight, $B(t)$, is simply the number of fish, $N(t)$, times weight per fish, $w(t)$. Biomass weight over time is illustrated in Figure 9.3; $t = t_0$ is the time of maximum biomass weight. Figure 9.3(b) presumably better illustrates a yearclass of farmed fish. The variations in yearclass weight can be the result of seasonal growth or variations in the mortality rate over time, or both.

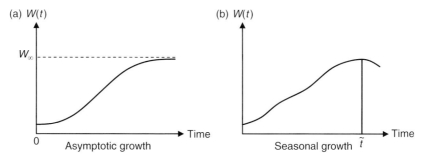

Figure 9.2   Weight curves.

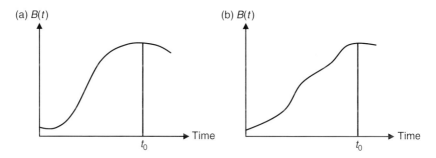

Figure 9.3   Biomass weight $B(t) = N(t) W(t)$.

9.2 Bioeconomic analysis

The biological model illustrates the changes in a yearclass of fish over time as a result of growth and natural mortality. Growth implies increased value for the fish farmer; mortality represents a loss. The purpose of bioeconomic analysis is to find the harvesting time that maximises the value of the yearclass to the fish farmer, within the given biological constraints.

The problem considered is an investment in fish and it is most easily understood in the context of investment theory. Simply put, one wishes to maximise the present value of an investment by determining the optimal time of harvesting; hence only costs that influence the cash flow the investment generates are relevant. In the present case this amounts to variable costs such as feeding and harvesting. To simplify the analysis it will be assumed that the number of recruits (fingerlings or smolts) is predetermined due to technological constraints.

The project analysed here concerns a one-time investment in a yearclass of fish. What happens after the lifetime of the project is not considered. This problem will be revisited below. The analysis will be developed gradually to focus on intuition, beginning with a situation with no costs. This will give an intuitive understanding that can be a useful reference for more

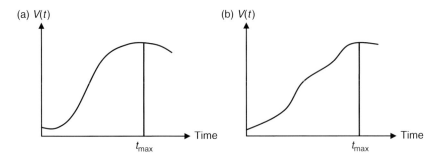

Figure 9.4   Biomass value $V(t) = p(w)B(t)$.

realistic models. Subsequently, different types of costs are introduced. Later, selective harvesting of fish is analysed.

9.2.1   Zero costs

The value of the yearclass is found by multiplying price by quantity:

$$V(t) = p(w)B(t) \tag{1}$$

where $V(t)$ is gross biomass value at time t and $p(w)$ is the price per kilogram of fish. The price is dependent on the size (weight) of the fish. Usually the price per kilogram is higher for large fish than for small fish, as shown in Chapter 6. The development in $V(t)$ over time, which is derived by multiplying yearclass weight (Figure 9.3a,b) by the price, is illustrated in Figure 9.4; $t = t_{max}$ is the time of the maximum biomass value. At time T the fish reaches its maximum individual weight, so that the value per fish is also greatest at that time. However, due to natural mortality, the yearclass will reach its maximum value earlier.

The fish farmer will plan to harvest at the time that maximises the present value of the biomass value[41] $\pi(t)$ as considered at the time of releasing the fish:

$$\max_{\{0 \le t \le T\}} \pi(t) = V(t)e^{-rt} \tag{2}$$

where r is the interest rate, and T the life expectancy of the fish or, alternatively, the time of sexual maturity. The harvesting time is the farmer's control variable.

At time t the farmer can harvest all fish and acquire an income of $V(t)$. Assuming that the farmer's alternative is to deposit the money in a bank account at a given interest rate of r, this will give a return on investment

---

[41]   The interest rate is here expressed as a continuous time interest rate. See Bjørndal (1990) for the relationship between continuous and discrete time interest rates.

equal to the interest paid on the value of the fish, $rV(t)$. Thus $rV(t)$ becomes the opportunity cost of the fish farmer. Alternatively, the farmer can refrain from harvesting at time t, allowing the fish to grow further, although some will die due to natural mortality. The change in biomass value over time will be the return on investment (in the form of fish, or capital in the sea) for keeping the fish in the sea. This can be denoted as $V'(t)$. At the optimal time of harvesting, $t = t^*$, the return on investment of the capital deposited in the sea equals the alternative one on land.

For the aquaculture operation to be commercially viable, the growth of the fish must for some periods of time be sufficiently large so that the investment of keeping the fish in the sea is better than puting the money in the bank, i.e. $V'(t) > rV(t)$. However, at some point, as the fish get larger, the growth rate declines and the natural mortality rate may also start to increase. This will eventually reverse the relationship and make $V'(t) < rV(t)$. Under this condition, the fish farmer will be better off by harvesting all the fish and depositing the money in the bank (or use it for some other purpose).

The following harvesting rules also illustrate the optimal rule.

(1)   Do not harvest the fish if the capital in the form of fish gives a better return on investment than one can obtain from a bank.
(2)   Harvest the fish if the capital in the form of fish gives a lower return on investment than one can obtain from a bank.

Hence, the optimal time to harvest the fish is when:

$$V'(t) = rV(t) \tag{3}$$

The optimal time to harvest the fish is illustrated in Figure 9.5, which depicts marginal revenue (MR) and marginal cost (MC) with respect to time. Marginal revenue is due to fish growth and associated price appreciation. As explained above, growth declines with time. For this reason, marginal revenue also declines with time. Marginal cost is composed of the alternative cost of capital (r) and the natural mortality rate (M), which is also a cost of keeping the fish in pens. The optimal time of harvesting, t^*, is when MR = MC.

If the growth rate increases, there will be an upward shift in the MR curve. If the price of fish depends on weight, increased price appreciation will also cause an upward shift in the MR curve.[42] In both cases it will be profitable for the farmer to keep the fish longer, as marginal cost remains unchanged. In other words, t^* has increased.

An increase in the discount rate causes an upward shift in the marginal cost curve. As the marginal revenue curve with respect to time is unchanged while

---

[42]   It must be noted that this refers to a situation where price increases with the weight of the fish. If price per kilogram is constant, a discrete change in the price of fish, e.g. from $4/kg to $5/kg, will not influence t^*. This will be explained in the Appendix.

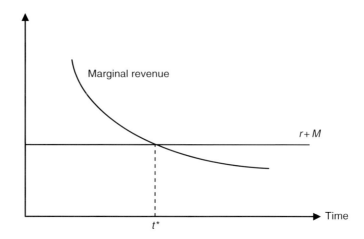

Figure 9.5   The optimal harvesting time t^*.

cost has increased, the optimal harvesting time is reduced. An increase in the natural mortality rate would have a similar effect on optimal harvesting.

9.2.2   Harvesting costs

Now the influence of costs on the harvesting time will be introduced. Assume that there is a fixed harvesting cost per kilogram fish of C_k, so that by harvesting all fish at time t the total harvesting costs will be $C_k B(t)$. Now $\{p(w) - C_k\}$ is the net price per kilogram the farmer receives when harvesting rather than $p(w)$. Assuming price depends on weight, price appreciation is now on the basis of net price $\{p(w) - C_k\}$ rather than on the basis of gross price $p(w)$. Thus the marginal revenue curve with respect to time has shifted upwards and the optimal harvesting time has increased. For price independent of weight, the optimal harvesting time is independent of the per unit harvesting cost.

Alternatively, the harvesting cost can be a fixed harvesting cost per fish of C_s, so that by harvesting all fish at time t the total harvesting costs will be $C_s N(t)$. The optimal harvesting time is also increased in this case, compared with the case of no harvesting costs.

9.2.3   Feed costs

So far growth has been regarded as a function of time. This ignores the relationship between feeding and growth. In fish farming this assumption is unrealistic and it will now be altered. The feed conversion ratio (f_t) is defined as follows:

$$f_t = \frac{F(t)}{w'(t)} \tag{4}$$

where $F(t)$ is the quantity of feed and $w'(t)$ is the change in the weight of the fish. The conversion ratio is thus the relation between the feed quantity and the growth of the fish. As a simplifying assumption, this factor is commonly set to be constant. The feed quantity per fish at time t is in general:

$$F(t) = f_t w'(t) = \text{conversion ratio} \times \text{growth} \qquad (4')$$

Note that the feed quantity varies over time according to the growth of the fish. Total feed quantity in any period can be found by multiplying by the number of fish. The fish are fed continuously over time. Total feed quantity is found by summing the feed quantity from the time the fish are released until time t.

From an economic point of view, feed costs are a concern since they reduce the cash flow. Let the price per unit (kg) of feed be C_f. Multiplying by total feed quantity gives the total feed costs at time t as $C_f F(t) N(t)$. The economic concern at any point in time is then the additional feed costs, if it is decided to keep the fish longer, relative to the additional revenue obtained by the increased biomass weight. Feed costs will cause an upward shift in the marginal cost curve in Figure 9.5. As a consequence, the optimal harvesting time is reduced. If there are seasonal variations in the feeding, as was shown for salmon in Chapter 2, the feed quantity and thereby the feed cost will be a non-linear function over time.

While the feed costs may imply that it is optimal to harvest the fish earlier than otherwise, the harvesting costs work in the opposite direction. The net result is an empirical question and varies from species to species. More generally, harvesting cost represent the way to introduce any one time cost, while the feed cost represent how any cost that occur throughout the production process can be specified. A more formal discussion and estimates of the optimal harvesting time that show the impact of different kinds of costs will be given in the appendix.

9.3   The rotation problem

The analysis of the optimal time for harvesting a yearclass did not consider that when the fish are actually harvested, space is made available for new fish. This is an important aspect as space (volume) in fish farms is limited by environmental considerations, available facilities or regulations. As growth slows down and the marginal value decreases over time, harvesting can make room for new fish that may grow faster and thereby yield a greater increase in value. Therefore, it is not sufficient to merely consider a single harvesting time. A sequence of such times must be determined. This is illustrated in Figure 9.6.

This problem may be clarified with reference to investment theory. So far, a one-time investment in fish has been considered without taking into

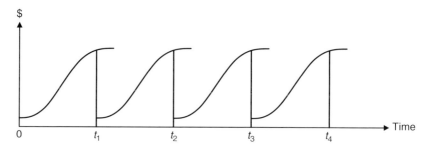

Figure 9.6   The rotation problem.

account what happens to the project when its lifetime has expired. When the fish are harvested, space is made available for releasing new fish. These new fish represent a new investment. Assuming the production capacity to be constant into the future, the problem represents an indefinite series of identical investments, not a one-time investment.

All biological and economic parameters are assumed to be constant over time. For this reason the length of cycles will be constant. The problem then is reduced to finding the optimal rotation time for a yearclass, i.e. the time of harvesting a yearclass and releasing the next. Consider a sequence of times:

$$t_1 < t_2 < t_3 < \dots$$

with the property that at time t_k the fish will be harvested and new fish will be released; $t = 0$ is the time of releasing the first yearclass. Rotation leads to a reduction in the optimal harvesting time. This is because the farmer can increase the total value of the production by replacing slower-growing old fish with faster-growing young fish, as shown in the Appendix.

It is here assumed that when one yearclass is harvested, the next one is released immediately. This implies that 'recruits' are available throughout the year. In reality this is not so for all species. Salmon spawn only at certain times (see Chapter 2). This determines when new smolts become available and makes the rotation problem more complicated. For some species such as turbot, it is possible to control the time of spawning so that fry are available for the whole year. The rotation problem is therefore relevant for farming these species.

Today, salmon are mainly harvested in the second year after being released. However, there is reason to believe that some market segments desire salmon of 2–3 kg aged less than 1 year. Harvesting in the first year is therefore a realistic alternative to harvesting in the second year. This is particularly true as breeding has increased growth rates. An economic analysis of this issue must be based on a discrete form of the model that is presented here.

Appendix: optimal harvesting of farmed fish

In this appendix, we will more formally show the model that was heuristically explained in the main text.

A biological model

This biological process can be described by an adapted Beverton–Holt model for a single yearclass:

$$N(0) = R \tag{A1}$$

$$\frac{dN}{dt} \equiv N'(t) = -M(t)N(t), \qquad 0 \le t \le T, \tag{A2}$$

$$N(t) = \mathrm{Re}^{-\int_0^t M(u)\,du} \tag{A3}$$

The variable t measures time from the release of fish as well as the age of fish. $N(t)$ is the number of fish at time t, while dN/dt denotes its time rate of change. At the outset, $t = 0$, R fish (recruits) are released (equation A1). Over time the number of fish in a yearclass changes due to natural mortality. This is given by the instantaneous mortality rate $M(t)$, which can vary over time (equation A2). $N(t)$ in equation A3 is the number of remaining fish at time t. The fish live until age T. One can alternatively consider this the time of sexual maturity.

If the mortality rate is assumed to be constant over time, i.e. $M(t) = M =$ constant, equations A2 and A3 simplify to the following:

$$\frac{dN}{dt} \equiv N'(t) = -MN(t), \qquad 0 \le t \le T \tag{A2$'$}$$

$$N(t) = \mathrm{Re}^{-M(t)} \tag{A3$'$}$$

Now let $w(t)$ symbolise the weight per fish at time t. The time rate of change in weight, i.e. the growth, is then $w'(t) \equiv dw/dt$. In general a **growth function** is expressed as follows:

$$w'(t) = g(w(t), N(t), F(t)) \tag{A4}$$

Growth is here expressed as a function of three variables: weight, number of fish (density) and feed quantity $F(t)$. It could be modelled as a function of biophysical variables such as temperature and light as well (see Chapter 2). These variables are here assumed to be a function of time.

In fish farming it is believed that excessive density can reduce growth, i.e. $g_N < 0$. However, if density does not influence growth, $g_N = 0$. For feeding to take place, it must have a positive impact on growth, i.e. $g_F > 0$. The individual fish will grow towards a maximum value either asymptotically or not. Define

$$w'(\tilde{t}) = 0$$

where $\tilde{t}$ is the time the fish reaches its maximum individual weight. The biomass weight for the yearclass at time t, $B(t)$, is defined as:

$$B(t) = N(t)\,w(t) = \mathrm{Re}^{-Mt}\,w(t) \tag{A5}$$

Here all fish are assumed to be of equal weight. Accordingly, the analysis is in terms of the 'representative' or average fish. As with individual fish, the yearclass will also reach a maximum size. Changes in biomass weight over time are given by:

$$
\begin{aligned}
B'(t) &= w'(t)N(t) + w(t)N'(t) \\
&= w'(t)N(t) - Mw(t)N(t) \\
&= \left[\frac{w'(t)}{w(t)} - M\right]B(t).
\end{aligned}
\tag{A6}
$$

$w'(t)/w(t)$ is the relative growth rate of the fish that presumably decreases over time, at least within the time interval that is relevant for harvesting. The following relations exist.

(1) $w'(t)/w(t) > (<) M$ implies that $B'(t) > (<) 0$, i.e. biomass weight increases (decreases).
(2) For $t = t_0$, where t_0 is defined by $w'(t_0)/w(t_0) = M$ so that $B'(t_0) = 0$, the biomass weight reaches its maximum. This occurs when the relative growth rate exactly equals the mortality rate.

Comparing the times for maximum individual weight and maximum yearclass weight reveals that $t_0 \le \tilde{t}$. The biomass of fish, $B(t)$, reaches its maximum before the individual fish, $w(t)$, reaches its maximum. This is due to the fact that when $B(t)$ reaches its maximum, the individual growth rate is still positive ($w'(t_0)/w(t_0) = M$), while the growth rate is reduced to zero when the individual fish reaches its maximum weight ($w'(\tilde{t}) = 0$).

Bioeconomic analysis

The purpose of this analysis is to find the harvesting time that maximises the present value of the net revenues from the yearclass, within the given biological constraints.

Zero costs

The economic analysis begins with a hypothetical example of zero costs. Define

$$V(t) = p(w)\, B(t) = p(w)\, \mathrm{Re}^{-Mt}\, w(t) \tag{A7}$$

where $V(t)$ is gross biomass value and $p(w)$ is the price per kilogram of fish.[43] The price is dependent on the size (weight) of the fish. Usually the price per kilogram is higher for large fish than for small fish ($p'(w) > 0$). The number of fish released (R) and the growth curve are also considered exogenous variables. The time $t = t_{max}$ is the time of the maximum biomass value, i.e. $V'(t_{max}) = 0$.

If $p'(w) = 0$, i.e. the price of fish is not dependent on its weight, then $t_{max} = t_0$. If $p'(w) > 0$, so that an increase in weight implies a higher price, then $t_{max} > t_0$. However, one does have $t_{max} < \tilde{t}$. Altogether, the following relations exist:

$$t_0 \leq t_{max} \leq \tilde{t}$$

The biomass value reaches its maximum value at the same time or later than the time of maximum weight of the yearclass, as price depends on weight or not. However, the yearclass reaches its maximum value earlier than the time of maximum individual weight of the fish. At time $\tilde{t}$ the fish reaches its maximum individual weight, so that the value per fish is also greatest at that time. Because of natural mortality, the yearclass will reach its maximum value earlier.

For the case of zero costs, the fish farmer will harvest at the time that maximises the present value of the biomass value:

$$\max_{\{0 \leq t \leq T\}} \pi(t) = V(t) e^{-rt}$$

Here, $\pi(t)$ is the present value of harvesting at time t, r is the interest rate, and T the life expectancy of the fish, or alternatively the time of sexual maturity. The harvesting time is the farmer's control variable. First-order conditions for an optimum:

$$\pi'(t) = V'(t)\, e^{-rt} - rV(t)\, e^{-rt} = 0$$

The optimal harvesting time, t^*, thus satisfies:

$$V'(t^*) = rV(t^*) \tag{A8}$$

---

[43]   As weight $w(t)$ is a function of the age of the fish, t, price is also a function of age.

or

$$\frac{V'(t^*)}{V(t^*)} = r \tag{A8'}$$

Additionally, the second-order conditions must be fulfilled.

The following harvesting rules also illustrate the optimal rule.

(1) Do not harvest the fish if $V'(t) > rV(t)$. In this case the capital in the form of fish gives a better return on investment than one can obtain from a bank.

(2) Harvest the fish if $V'(t) \leq rV(t)$.

In the 'traditional' Beverton–Holt model the harvesting time is determined unambiguously, although this is not always the case. This question will not be pursued here.

Subsequently it shall be assumed that the optimal harvesting time, t^*, is unambiguously determined. For $r > 0$, $t^* < t_{max}$. This implies that the optimal time of harvesting is less than the time that yields maximum biomass value (t_{max}). This is a result of discounting future incomes. If there is no discounting, i.e. $r = 0$, the optimal harvesting time is given by the condition:

$$V'(t^*) = 0$$

In this situation, $t^* = t_{max}$ and $V(0) = V(t_{max})$. In this case harvesting will take place when the nominal biomass value reaches its maximum. Whether it is optimal to harvest the fish before or after the time of maximum biomass weight, t_0, depends on the values of the parameters.

It is possible to acquire a better understanding of the harvesting rule by evaluating changes in the separate elements in the biomass value:

$$V'(t) = p'(w)w'(t)\mathrm{Re}^{-Mt} w(t)$$
$$- Mp(w)\mathrm{Re}^{-Mt} w(t)$$
$$+ w'(t)p(w)\mathrm{Re}^{-Mt} \tag{A9}$$

The first component of this expression takes into account the increment in value due to increased weight ($w'(t)$) that causes the price to rise ($p'(w) > 0$). The second component represents the economic loss resulting from the natural mortality of fish; the third component shows the increased value due to growth. The expression for $V'(t)$ can be rewritten as:

$$V'(t) = \left\{\frac{p'(w)}{p(w)}w'(t) - M + \frac{w'(t)}{w(t)}\right\}V(t) \tag{A9'}$$

The three components in brackets express the price appreciation due to growth, the natural mortality rate and the growth rate.

The rule for optimal harvesting says that the fish must be harvested when the marginal increase in the value of the 'natural' capital (i.e. fish in the sea) exactly equals the opportunity cost:

$$V'(t^*) = \left\{ \frac{p'(w)}{p(w)} w'(t^*) - M + \frac{w'(t^*)}{w(t^*)} \right\} V(t^*) = rV(t^*) \qquad (A10)$$

which can be rewritten as:

$$\frac{p'(w)}{p(w)} w'(t^*) + \frac{w'(t^*)}{w(t^*)} = r + M \qquad (A10')$$

The marginal revenue per fish with respect to time is $[p'(w)w'(t)w(t) + p(w)w'(t)]$. Marginal user cost per fish is $|r + M| p(w)w(t)$. Equating marginal revenue and marginal user cost per fish and dividing by the value per fish $(p(w)w(t))$ yields equation A10'.

The optimal harvesting time, t^*, was shown in Figure 9.5. The marginal revenue curve $\left\{ \frac{p'(w)}{p(w)} w'(t) + \frac{w'(t)}{w(t)} \right\}$ is declining over time, while the marginal cost $\{r + M\}$ is constant and thus represented by a horizontal line.[44] t^* is given by the intersection of the two curves. The fish must not be harvested when the natural capital (fish in the sea) gives a better return than the opportunity cost.

An increase in the growth rate will cause an upward shift in the marginal revenue curve. The same is the case if the rate of price appreciation increases. In both cases this would cause an increase in the optimal harvesting time.

An increase in the discount rate causes an upward shift in the marginal cost curve, and the optimal harvesting time is reduced. An increase in the natural mortality rate would have a similar effect on optimal harvesting.

Harvesting costs

Harvesting cost per quantity unit (kg)

Assume that there is a fixed harvesting cost per kilogram fish of C_k. The farmer faces the following maximisation problem:

$$\max_{\{0 < t \leq T\}} \pi(t) = \left\{ p(w)B(t) - C_k B(t) \right\} e^{-rt}$$
$$= \left\{ p(w) - C_k \right\} B(t) e^{-rt}$$

---

[44]   It is interesting to note that mathematically r and M play identical 'roles' in the solution to this problem.

Now $\{p(w) - C_k\}$ is the net price per kilogram the farmer receives when harvesting. The first-order condition for a maximum is given by:

$$\pi'(t) = p'(w)w'(t)B(t)e^{-rt} + \{p(w) - C_k\}B'(t)e^{-rt} - r\{p(w) - C_k\}B(t)e^{-rt} = 0$$

Through simplification, an implicit expression for the optimal harvesting time is derived:

$$\frac{p'(w)w'(t)}{p(w) - C_k} + \frac{w'(t^*)}{w(t^*)} = r + M \qquad (A11)$$

Assuming $p'(w) > 0$, one notices that the price appreciation term is different compared with equation A10′. Price appreciation is now on the basis of net price $\{p(w) - C_k\}$ rather than on the basis of gross price $p(w)$ as in equation A10′. Thus the marginal revenue curve with respect to time has shifted upwards and the optimal harvesting time has increased. For the case of $p'(w) = 0$, however, the price appreciation term vanishes. The optimal harvesting time is then independent of the per unit harvesting cost.

Harvesting cost per fish

Assume that there is a fixed harvesting cost per fish of C_s. The maximisation problem becomes:

$$\max_{\{0 < t \leq T\}} \pi(t) = \{V(t) - C_s N(t)\}e^{-rt} = \{p(w)w(t) - C_s\}Re^{-(M+r)t}$$

The first-order condition is given by

$$\pi'(t) = \{p'(w)w(t) + p(w)\}w'(t)Re^{-(M+r)t}$$
$$-(M + r)\{p(w)w(t) - C_s\}Re^{-(M+r)t} = 0$$

Through simplification, the following expression for the optimal harvesting time is derived:

$$\frac{p'(w)}{p(w)}w'(t^*) + \frac{w'(t^*)}{w(t^*)} = [r + M]\left\{\frac{p(w)w(t^*) - C_s}{p(w)w(t^*)}\right\} \qquad (A12)$$

Compared with equation A10′, one notices that the marginal revenue by not harvesting the fish is unaltered. However, the marginal cost has changed as a result of the introduction of harvesting costs. $p(w)w(t)$ is the gross value per fish at time t, whereas $\{p(w)w(t) - C_s\}$ constitutes the net value. The new expression on the right-hand side of the equation is the net value as a fraction of the gross value, which is less than 1. Multiplied by $r + M$, this gives

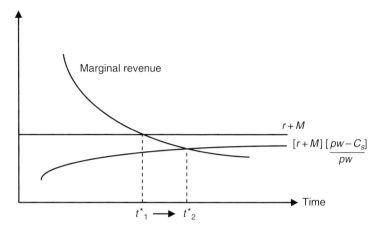

Figure A1 The optimal harvesting time with and without a harvesting cost per fish.

the marginal cost with respect to time, which in this way is reduced compared with the original formulation of the problem. As before one can deduce a harvesting rule: the fish should not be harvested as long as the natural capital (i.e. the fish) yields a better return than one can obtain alternatively.

The optimal harvesting time is illustrated in Figure A1. Without harvesting costs $t^* = t_1^*$, and with harvesting costs $t^* = t_2^*$ with $t_2^* > t_1^*$. As shown, harvesting costs imply that one must harvest later than otherwise. By postponing the harvesting a while, the discounted value of the harvesting costs will be reduced as a consequence of natural mortality. Discounting by itself also reduces the present value of harvesting costs. Further, an increase in the harvesting cost causes a downward shift in the marginal cost curve, thereby increasing the optimal harvesting time.

Feed costs

The conversion ratio (f_t) is defined as follows:

$$f_t = \frac{F(t)}{w'(t)} \tag{A13}$$

where $F(t)$ is the quantity of feed. The conversion ratio is thus the relation between the feed quantity and the growth of the fish. As a simplifying assumption, this factor is commonly set to be constant. The feed quantity per fish at time t is then:

$$F(t) = f_t\, w'(t) \tag{A13'}$$

Note that the feed quantity varies over time according to the growth of the fish. Total feed quantity at a given time is:

$$F(t)\,N(t) = F(t)\,\mathrm{Re}^{-Mt} = f_t\,w'(t)\,\mathrm{Re}^{-Mt} \qquad (A14)$$

The equation takes into account the loss of fish resulting from natural mortality.

The fish are fed continuously. Total feed quantity (SF_t) is found by summing the feed quantity from the time the fish are released until time t. This can be done by integrating equation A14:

$$SF_t = \int_0^t F(u)\mathrm{Re}^{-Mu}\,du \qquad (A15)$$

Let the price per unit (kg) of feed be C_f, which is constant over time. Multiplied by SF_t (equation A15) this gives the total feed costs at time t. Discounted back to the time of releasing the fish, $t = 0$, one obtains:

$$\text{Present value of feeding costs} = \int_0^t C_f F(u)\mathrm{Re}^{-Mu}e^{-ru}\,du$$

The farmer's maximisation problem is as follows:

$$\max_{\{0 \le t \le T\}} \pi(t) = V(t)e^{-rt} - \int_0^t C_f F(u)\mathrm{Re}^{-(M+r)u}\,du$$

The first-order condition for profit maximisation is then:

$$\pi'(t) = V'(t)\,e^{-rt} - rV(t)\,e^{-rt} - C_f F(t)\,\mathrm{Re}^{-(M+r)t} = 0$$

By use of equations A7 and A9 and through simplification, this can be rewritten as:

$$\frac{p'(w)}{p(w)}w'(t^*) + \frac{w'(t^*)}{w(t^*)} = r + M + \frac{C_f F(t^*)}{p(w)w(t^*)} \qquad (A16)$$

In comparison with the original first-order condition for profit maximisation (equation A10′), the marginal feed costs must now be included in the opportunity cost. $C_f F(t)$ are the feed costs per fish at time t and $p(w)w(t)$ is the value of the fish. $C_f F(t)/p(w)w(t)$ is thus the relative feed cost, which combined with the interest rate and the rate of natural mortality constitutes the cost of not harvesting the fish at time t. Determination of the optimal harvesting time is illustrated in Figure A2. In this situation the marginal

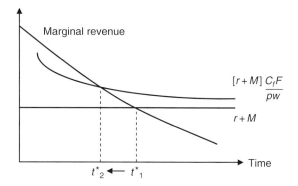

Figure A2   The optimal harvesting time with and without feed costs.

revenue curve with respect to time is as before, but the marginal cost has increased as a result of the feed costs. This implies harvesting the fish earlier than before.

Feed and harvesting costs

When all costs are included, the following rule for the optimal harvesting time is derived:

$$\frac{p'(w)w'(t)}{p(w)-C_k} + \frac{w'(t^*)}{w(t^*)} = (r+M)\left\{\frac{p(w)w(t^*)-C_s}{p(w)w(t^*)}\right\} + \frac{C_fF(t^*)}{p(w)w(t^*)} \qquad (A17)$$

If the value of the fish are insured at a premium of k, so that $kV(t)$ is the insurance premium paid at any point in time, k would enter the first bracket on the right-hand side. Mathematically, it would be identical to r and M.

While the feed costs imply that it is optimal to harvest the fish earlier than otherwise, the harvesting costs work in the opposite direction. The net result is an empirical question and varies from species to species (see below).

The rotation problem

This problem may be clarified with reference to investment theory. So far, a one-time investment in fish has been considered without taking into account what happens to the project when its lifetime has expired. When the fish are harvested, space is made available for releasing new fish. These then represent a new investment. Assuming production capacity to be constant into the future, the problem represents an indefinite series of identical investments and not a one-time investment. This aspect will now be included in the analysis.

All parameters are assumed to be constant over time. The problem then is reduced to finding the optimal rotation time for a yearclass, i.e. the time of harvesting a yearclass and releasing the next.

Consider a sequence of times:

$$t_1 < t_2 < t_3 < \ldots$$

with the property that at time t_k the fish will be harvested and new fish will be released; $t = 0$ is the time of releasing the first yearclass. This is the same as the first investment in fish. The present value (PV) at time $t = 0$ of all future incomes is given by:

$$PV = V(t_1)e^{-rt1} + V(t_2)e^{-rt2} + V(t_3)e^{-rt3} + \ldots \qquad (A18)$$

where r is the interest rate. Rotation is found to imply a reduction in the optimal harvesting time. This is because the farmer can replace slower-growing old fish with faster-growing young fish.

The maximisation of equation A18 with respect to $t_1, t_2, t_3 \ldots$ appears to be arduous. What simplifies the problem is that as all parameter values are constant over time, the rotation periods are of equal length, i.e.

$$t_k = kt \quad k = 1,2,3 \ldots \qquad (A19)$$

where t is the rotation time. Inserting equation A19 into A18 gives the following maximisation problem for the fish farmer (see summation formula below):

$$\text{Max } \pi(t) = V(t)e^{-rt} + V(t)e^{-2rt} + V(t)e^{-3rt} + V(t)e^{-4rt} \ldots \qquad (A20)$$

This can be simplified by help of the following manipulation:

(i) $\text{Max } \pi(t) = V(t)e^{-rt} + V(t)e^{-2rt} + V(t)e^{-3rt} + V(t)e^{-4rt} + \ldots$

Multiply both sides by e^{rt}:

(ii) $e^{rt}\pi(t) = V(t) + V(t)e^{-rt} + V(t)e^{-2rt} + V(t)e^{-3rt} + \ldots$

Subtracting (i) from (ii):

$$\pi(t)\,(e^{rt} - 1) = V(t)$$

This gives

$$\pi(t) = \frac{V(t)}{e^{rt} - 1}$$

The first-order condition for a maximum is given by:

$$\pi'(t) = \frac{V'(t)[e^{rt}-1]-re^{rt}V(t)}{[e^{rt}-1]^2} = 0.$$

This can be simplified to

$$\frac{V'(t^*)}{V(t^*)} = \frac{r}{1-e^{-rt^*}} \qquad (A21)$$

t^* is now to be understood as the **optimal rotation time**. Equation A21 is known as the Faustmann rule after the German forester Martin Faustmann, who was the first to solve the optimal rotation problem in forestry. By comparing this expression with the formula for optimal harvesting time for a single yearclass (equation A8'), it is seen that the left-hand side is the same, while there is a new component on the right-hand side. As $(1 - e^{-rt}) < 1$, rotation implies a reduction in the optimal harvesting time compared with no rotation.

Equation A21 can be rewritten as:

$$V'(t^*) = rV(t^*) + r\frac{V(t^*)}{e^{rt^*}-1} \qquad (A21')$$

As before, $V'(t)$ is the change in biomass value over time, whereas $rV(t)$ is the opportunity cost of the biomass value. The expression

$$\frac{V(t)}{e^{rt}-1}$$

is the present value of future cash flows and is the component that takes the rotation aspect into account. It gives the value of the fish farm which, multiplied by the interest rate, determines the opportunity cost of the fish farm. This value is a result of the fact that farm size is limited.

A special case occurs when the interest rate is zero. By use of l'Hopital's rule, the right-hand side of equation A21 becomes:

$$\lim_{r \Rightarrow 0} \frac{r}{1-e^{-rt}} = \frac{1}{t}$$

so that equation A21 becomes:

$$\frac{V'(t^*)}{V(t^*)} = \frac{1}{t^*}, \qquad r = 0 \qquad (A22)$$

or

$$\frac{V(t^*)}{t^*} = V'(t^*), \qquad r = 0 \tag{A22'}$$

In this case fish should be harvested when the time change in the biomass value ($V'(t)$) equals the average biomass value ($V(t)/t$). This means that one maximises annual biomass value $V(t)/t$.

Optimal harvesting: examples

This section provides examples of optimal harvesting for salmon. Previously the effect of different costs on harvesting time has been analysed. Now, through numerical examples, some quantitative results are obtained. Examples of weight curves are given by:

(1) $w(t) = w_\infty \left(a - be^{-ct}\right)^3$

(2) $w(t) = e^{a - b/t}$

(3) $w(t) = w_0 + a_1 t + a_2 t^2 + a_3 t^3$

(4) $w(t) = \dfrac{w_\infty}{1 + ae^{-bt}}$

The first function is von Bertalanffy's weight function, where w_∞ is the maximum individual weight fish will reach asymptotically over time. This function was originally developed for wild fisheries. In the second case, the fish will grow towards an asymptotic value (e^a), while the third is given by a third-degree polynomial. The last case is given by a logistic function where the fish also grow towards an asymptote (w_∞). The functional form one uses in applied work depends on which fits the data best.

Weight curves vary from species to species and from location to location, depending on several factors. The results represented here must therefore only be considered as examples. In spite of this, there is reason to believe that certain qualitative characteristics of the results may be generalised.

We will here use a weight curve estimated as a polynomial of time. The following weight curve has been estimated, based on growth observations for salmon.[45]

$$w(t) = 5.72t^2 - 2.08t^3$$

$w(t)$ is the weight of the fish at time t, measured in years from the time they were released. For this set of data a third-degree polynomial, with the constant and the first-degree component set equal to zero, gave good results.

The estimated growth curve is illustrated in Figure A3. The following assumptions have been made about natural mortality and price:

---

[45]   The weight curve is estimated using the ordinary least squares method. t statistics for the two parameter estimates are 32.81 and −12.52 respectively.

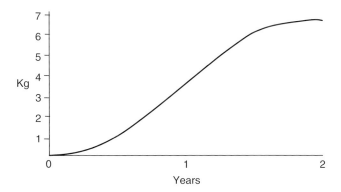

Figure A3   Weight curve for salmon.

$$M = 0.1$$
$$p(w) = 16.1 + 0.5w(t)$$

This implies that approximately 10% of the fish will die naturally in 1 year.[46] The price is in Norwegian kroner (NOK) per kilogram.[47] With these parameter values one obtains

$$\tilde{t} = 1.833 \text{ years}$$

$$w(\tilde{t}) = 6.41 \text{ kg}$$

$$t_0 = 1.776 \text{ years}$$

$$t_{max} = 1.784 \text{ years}$$

According to the weight curve, the fish reaches its maximum weight, 6.41 kg, at 1.833 years ($\tilde{t}$) after being released. The yearclass, however, reaches its maximum weight after 1.776 years (t_0). As the fish price increases with the weight of fish, the maximum biomass value is reached later, here after 1.784 years.

The following assumptions have been made about costs:

- Feed price per kg (C_f) = NOK7.00
- Conversion ratio (f_t) = 1.10
- Harvesting cost per fish (C_s) = NOK15.00.

Table 9.1 gives the optimal time of harvesting salmon for different cost alternatives and interest rates. In the zero cost alternative, fish are to be harvested at the time of maximum biomass value $V(t)$, i.e. $t^* = t_{max}$ for $r = 0$.

---

[46]   This parameter value gives a total natural mortality of more than 20% over the lifetime of the fish. This is not unreasonable.

[47]   This price function is a linear approximation to ex-farm salmon prices.

Table 9.1 Optimal harvesting time (t^*) in years for salmon and optimal rotation period (t^*).

Interest rate (r)	t^* (zero costs)	t^* (harvesting costs)	t^* (feed costs)	t^* (feed and harvesting costs)
0.00	1.784	1.790	1.726	1.735
0.05	1.759	1.768	1.684	1.700
0.10	1.732	1.745	1.642	1.663
0.15	1.706	1.722	1.599	1.626
0.20	1.679	1.699	1.554	1.588

Table 9.2 Optimal rotation period (t^*) zero costs.

Interest rate (r)	t^*
0	1.373
0.05	1.352
0.10	1.331
0.15	1.310
0.20	1.287

There are two conclusions to draw from the results. First, the optimal harvesting time is relatively insensitive to changes in the interest rate. An increase in r from 0.1 to 0.2 reduces t^* by about half a month. Second, the harvesting time is only to a small extent influenced by variable costs. Note that the harvesting costs imply an increase in t^*, but feed costs imply a reduction in t^*. This corresponds to the analytical results derived above. In the case considered here, the feed costs dominate. The net result is that the costs imply a decrease in t^* compared to a situation with no costs.

A change in the natural mortality rate (M) will have an identical effect on the optimal harvesting time as an equivalent change in the interest rate. This is clearly indicated in equation A10'.

We also consider optimal rotation for the zero cost case (Table 9.2). When rotation is considered, this has a very large impact on harvesting case. For r = 0 and zero costs, rotation implies a reduction in harvesting time from 1.784 to 1.373 years or almost 5 months. A similar result is found for other discount rates. This is a substantial impact which indicates that, if rotation is possible, it will have important consequences for the optimal production plan.

Bibliography

The optimal harvesting problem for farmed fish is not qualitatively different from other forms of animal production, or other resource industries with a biological growth function and a limited production space. All harvest models can therefore be traced back to Faustmann's seminal work on optimal tree rotation.

As salmon and shrimp have been the most successful and valuable aquaculture species, it is not surprising that the most attention has been given to them. Cacho (1997) provides a review of this literature. Karp *et al.* (1986) and Leung and Shang (1989), among the earliest studies focusing on aquaculture, consider the problem of determining optimal harvest and restocking times and levels for farmed shrimp.

Bjørndal (1988) developed the first optimal harvesting models for aquaculture based on the forestry literature. Several authors have extended Bjørndal's model to emphasise specific aspects of the problem. Arnason (1992) introduced dynamic behaviour and presented a general comparative dynamic analysis. He also introduced feeding as a decision variable. Heaps (1993) dealt with density-independent growth, whereas Heaps (1995) allowed for density-dependent growth and looked at the culling of farmed fish. Mistiaen and Strand (1998) demonstrated general solutions for optimal feeding schedules and harvesting times under conditions of piecewise-continuous, weight-dependent prices, while Guttormsen (2008) allows for restricted smolt release as well as different relative prices between weight classes. Rizzo and Spagnolo (1996) provided one of the first applications for sea bass.

10 Production Planning in a Salmon Farm

While the model in Chapter 9 gives a useful insight into optimal harvesting in a salmon farm, it also excludes a number of real world phenomena that must be taken into account. This includes the fact that continous time operations are not realistic in real life, as well as issues such as purchasing of input factors, financing and acquiring farm licences. This and the next chapter analyse production planning in salmon farms from a more practical perspective. The analysis makes use of a discrete time model that is updated once a month with respect to important variables such as the number of fish, growth, feeding and mortality. It is based on the continuous time model expounded in Chapter 9, and the same basic method is used. However, the monthly model is easier to use in actual fish farm operations.

In this chapter we will look at the short-run decisions related to a single release of fish on an existing farm after the smolts have been purchased. In Chapter 11 we will analyse the investment decision in the farm, and also the costs that are not directly related to a single release of fish. Therefore, the focus here is on feeding the released fish and determining when to harvest. The principles of cash flow analysis are discussed in section 10.1, as they constitute the basis for the analysis. In section 10.2 the biological characterics of smolt release and biomass growth are provided. In section 10.3 we discuss the sales revenue, in section 10.4 feed costs and in section 10.5 the net present value of the investment. In section 10.6 we allow the harvesting to be spread over a longer period of time.

The primary intent of this chapter is to give a practical illustration of the method employed in optimal production planning. Because of the many calculations involved, sensitivity analyses for changes in parameter values are not carried out. For actual planning in fish farms, computer models are used, and several software packages are commercially available that allow sensitivity analysis to be carried out in a straightforward manner. The data used will be based on values for farmers of Atlantic salmon in western Norway and their role is primarily to illustrate the decision process. The data represent the situation faced by producers in Norway in the current cycle. The data are meant to be representative for a typical firm at mid-cycle. All monetary values are expressed in nominal Norwegian kroner (NOK).

10.1   Cash flow analysis

The starting point in this section is a fully developed fish farm. Only one release of fish is considered. In this way all the major results are derived at an intuitive level. Any investment analysis is in terms of the cash flow the investment generates with respect to both revenues and expenditures. In this case it is based on monthly periods. Let V_t represent revenue when harvesting in month t, C_t expenditures in month t, and r the monthly interest rate.[48]

Fish are released at time zero, i.e. $t = 0$. It is assumed that there is a limit on how long the fish farmer can keep the fish before harvest. This limit will in practice be determined by when the next batch is to be released or when the fish reach sexual maturity. Assuming this limit occurs after 19 months, the following alternatives exist with respect to harvesting.

	$t = 0$	$t = 1$	$t = 2...$	$t = 19$
Sales revenue	V_0	V_1	V_2	V_{19}
Expenditures	C_0	C_1	C_2	C_{19}
Net cash flow	KS_0	KS_1	KS_2	KS_{19}

The problem facing the fish farmer is harvesting the fish in the month that maximises the present value of the investment. Assuming all fish will be harvested in the same month, there are 20 possible investment alternatives, i.e. harvesting in the months zero to 19. In fact harvesting is only realistic for, say the last 12 months, as growth is fairly fast for the small fish, and their market value is low. It is also possible to spread the harvest over several months. This option will be discussed below.

KS_t, $t = 0,1, \ldots 19$, shows the net cash flow in nominal values if harvesting takes place in this month, i.e. sales revenue minus expenditures in the month of harvesting. In addition, expenditures in previous months must be taken into account.[49] In order to find the present value (PV) of the investment, cash flows must be discounted back to the time of releasing the fish. There are the following alternatives for the investment in fish.

---

[48]   The interest rate measures the opportunity cost of capital. If the capital were deposited in a bank account were it not invested in the fish farm, the bank interest rate would be the opportunity cost to the investor. Normally, the opportunity cost would be the expected return on other investment opportunities available to the investor.

[49]   In principle, revenues in months other than the month of harvesting should also be taken into account. In this example, however, there are sales revenues only in the month of harvesting, so other months can be disregarded.

Harvest **t = 0**

$$PV_0 = KS_0 = V_0 - C_0$$

Harvest **t = 1**

$$PV_1 = \frac{KS_1}{1+r} - C_0 = \frac{V_1}{1+r} - \left\{ C_0 + \frac{C_1}{1+r} \right\}$$

Harvest **t = 2**

$$PV_2 = \frac{KS_2}{(1+r)^2} - \left\{ C_0 + \frac{C_1}{1+r} \right\} = \frac{V_2}{(1+r)^2} - \left\{ C_0 + \frac{C_1}{1+r} + \frac{C_2}{(1+r)^2} \right\}$$

Harvest **t = n; n = 0, ..., 19**

$$PV_n = \frac{KS_n}{(1+r)^n} - \left\{ C_0 + \frac{C_1}{1+r} + \frac{C_2}{(1+r)^2} + ... + \frac{C_{n-1}}{(1+r)^{n-1}} \right\}$$

$$= \frac{KS_n}{(1+r)^n} \qquad - \qquad \sum_{t=0}^{n-1} \frac{C_t}{(1+r)^t}$$

$$= \frac{V_n}{(1+r)^n} \qquad - \qquad \sum_{t=0}^{n} \frac{C_t}{(1+r)^t}$$

$$= \text{Present value} \quad - \quad \text{Present value}$$
$$\text{sales revenue} \qquad \text{expenditures}$$

As can be seen, it can be useful to discount the cash flows for revenues and expenditures separately. By calculating the present value of harvesting in all possible months, the month that maximises the present value can be found. This is analogous to the maximisation in the continuous time analysis in the previous chapter.

In many farms, the maximum biomass available at the farm is constrained by the physical size of the farm, by regulation, or because density limits growth. When a capacity restriction becomes binding, it means that some fish must be harvested earlier than optimal because of the constraint. As this fish will be sold, it provides revenues in earlier periods, although it will also increase costs in those periods since the firm incurs harvesting costs. In this case, the present value of the cash flow will be given as:

$$PV_n = \sum_{n=n-i}^{n} \frac{KS_n}{(1+r)^n} - \sum_{t=0}^{n-i-1} \frac{C_t}{(1+r)^t}$$

$$= \sum_{n=n-i}^{n} \frac{V_n}{(1+r)^n} - \sum_{t=0}^{n} \frac{C_t}{(1+r)^t}$$

The first expression indicates that in the first $n - i - 1$ period there will only be costs, while in the remaining $n - i$ periods there will be a net cash flow that in any period can be positive or negative. In the second expression, revenue and cost are separated. While costs are incurred in all n periods, revenues start in period $n - i$, the first period of harvesting, and then continue until all fish are harvested. As such, the main problem is when to harvest the remaining fish. At any point in time this will be determined by maximising net present value over the remaining lifespan of the fish.

10.2   Smolt release and biomass growth

The reminder of this chapter will be concerned with an example of optimal harvesting of one smolt release. The reference point for the analysis is then the time of smolt release. The farmer's investment in smolts, and thereby the number of fish released, is considered as given. The objective is to find the harvest time that maximises the present value of the investment in smolts. The following five assumptions constitute the basis of the analysis:

(1)   The plant is fully developed. All fixed costs relating to plant operation are disregarded.
(2)   Labour is here regarded as a fixed cost.
(3)   Smolts have been purchased and released. Therefore only the harvesting time with respect to the number of smolts already released is of interest.
(4)   What happens after the yearclass has been harvested (rotation) is not considered, i.e. only a one-time investment is considered.
(5)   Credit is not a limiting constraint with respect to operations and taxes are disregarded.

Given these assumptions, the farmer's objective is to maximise the present value of cash flows from the investment represented by the smolt release. Hence, only variables directly affected by the investment are of relevance.

In addition to biological variables, we will show the development of feed and harvesting costs as well as revenues and net earnings for a given set of prices and interest rate. Harvesting costs include hiring additional labour during harvesting when this is done in-house, or alternatively payment to the slaughtering firm when harvesting is outsourced. The analysis can easily be extended to include other variable costs. However, it is the principal differences between the various types of costs that are important. Feed costs are incurred every month, while harvesting costs are only incurred in the month(s) of actual harvesting. Other types of costs will have a similar structure. It is assumed that revenues are obtained in the same month as harvesting takes place and that expenditures are paid in the month incurred. Hence problems relating to periodising revenues and expenditures are disregarded.

This analysis is based on the following assumptions concerning plant operations and development of the yearclass:

- In the month of May, year 0, 500 000 salmon smolts are released. This means the month of May represents $t = 0$.
- Natural mortality is 0.5% per month throughout year 0. In year 1 it increases to 1% in March and April, 2% in May and June, 3% in July, 4% in August, 6% in September, 8% in October, 11% in November and 12% in December.[50]
- The fish die or are harvested at the end of the month. The fish are normally starved before harvest, and during this period lose some weight. Assume this loss is 1% of what they would weigh if fully fed for for this period.[51]

Table 10.1 and Figure 10.1 illustrate the development of the yearclass over time. As noted, the fish are released in May. The second and third columns show the weight per fish (w_t) at the start of each month and the weight increase per fish $(w_{t+1} - w_t)$ during the month. The weight per fish is derived from a growth function of the type discussed in Chapter 2. Weight is continuously increasing over time, although growth is slower during winter. For the growth function used, a fish would not reach maximum weight until the end of year 3. However, as this possibility has no practical relevance, the table ends in January in year 2.

The number of fish, N_t, is given in the fourth column and starts with the half million smolts that were released. A natural mortality rate of 0.5% per month is assumed early on, so there are losses every month. In the spring of year 1 the mortality rate starts to increase as sexual maturity becomes an issue, and the number of fish decreases more rapidly. In the final column the biomass weight, B_t, measured in tonnes at the start of the month is shown. This reaches a maximum of 2503 tonnes in October of year 1. After this, natural mortality is greater than the weight increase so that the net change in biomass weight is negative.

---

[50] Strictly speaking, mortality does not describe a sexually mature fish. However, such fish lose their value as they cannot be marketed, and therefore cannot be regarded as commercially perished.

[51] Since feeding depends on temperature, this period will depend on temperature. Normally, the fish are not fed for about 50 day-degrees. Thereafter, the fish are transferred to the harvesting plant, where they are kept in a holding pen until about 70 day-degrees. We simplify by setting weight loss to 1%.

Table 10.1   Development of biomass.

Month	Weight per fish (kg) w_t	Weight increase (kg) $w_{t+1} - w_t$	Number of fish N_t	Biomass weight (tonnes) $B_t = N_t w_t / 1000$
Year 0				
May	0.106	0.065	500 000	53
June	0.171	0.124	497 500	85
July	0.294	0.229	495 013	146
August	0.523	0.312	492 537	258
September	0.835	0.368	490 075	409
October	1.202	0.367	487 624	586
November	1.569	0.365	485 186	761
December	1.934	0.361	482 760	934
Year 1				
January	2.295	0.307	476 967	1095
February	2.602	0.294	470 290	1224
March	2.896	0.292	463 235	1342
April	3.189	0.381	456 287	1455
May	3.570	0.479	447 161	1596
June	4.049	0.642	435 982	1765
July	4.691	0.788	422 031	1980
August	5.479	0.872	403 039	2208
September	6.351	0.908	376 842	2393
October	7.259	0.869	344 810	2503
November	8.127	0.862	305 157	2480
December	8.990	0.845	267 012	2400
Year 2				
January	9.835	0.731	230 966	2272

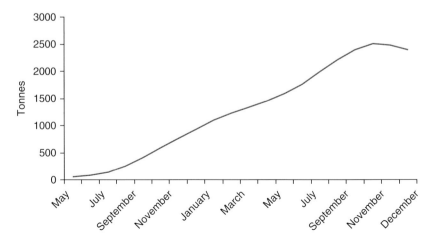

Figure 10.1   Biomass development, May year 0 to December year 1.

Table 10.2   Development of biomass value. Monthly discount rate 0.65%.

Month	Price per kg (NOK)	Biomass value (nominal) (million NOK) $V_t = p(w)(0.99)B_{t+1}$	Biomass value (discounted) (million NOK) $V_t/(1 + r)^t$
Year 1			
January	18.59	22.5	21.4
February	18.71	24.9	23.4
March	18.98	27.3	25.6
April	19.02	30.1	28.0
May	19.19	33.5	31.0
June	19.23	37.7	34.6
July	19.26	42.1	38.5
August	19.30	45.7	41.5
September	19.30	47.8	43.1
October	19.32	47.4	42.5
November	19.34	46.0	40.9
December	19.35	43.5	38.5

10.3   Sales revenue

It has been assumed that all fish are harvested at the end of the month. However, as noted above the fish are not fed during the last days before harvest and there is a weight loss of 1% assumed relative to the end weight. This must be taken into account when computing the biomass value. Thus, the actual sales volume will equal 99% of the biomass weight:

$$V_t = p(w)(0.99)B_{t+1}, t = 0,1,2,\ldots,21$$

where B_{t+1} is biomass weight in month $t + 1$ (as this is the number of fish and the biomass at the end of month t), $p(w)$ the price per kilogram and the coefficient of 0.99 corrects for the weight loss due to starving of the fish prior to harvesting.

Table 10.2 shows price per kilogram and two measures of biomass value, nominal and discounted, on a monthly basis for year 1.[52] In the second column the price is shown. The price per kilogram depends on the weight of the fish, and the fish are divided into different size groups with price generally increasing with weight. The biomass value at a point in time is found by finding the size distribution of the fish and calculating the value while

---

[52]   We do not report any information for year 0, as the fish are then so small and the growth so high that no economic decision would lead to harvesting of the fish.

taking into account the fact that prices vary with size. This has been done for the case considered here.[53] The price column in Table 10.2 represents average price for each month in question.

An alternative to this procedure is to use the average weight of the fish as shown in Table 10.1, and then to use the price per kilogram of fish in the respective size categories. The disadvantage of this approach is that when the average weight from one month to the next moves from one size category to another, for example from 2.7 to 3.1 kg (i.e. moving average size from the 2–3 kg range to the 3–4 kg range), it might result in an overstatement of the increase in value. When the mean weight equals 3.1 kg, over 30% of the fish are in the less than 3 kg category, while the rest belong to the 3 kg and over category. In this example the price per kilogram for 3–4 kg fish will not give a precise estimate if it is used to assess biomass value.

With the assumptions made here, average price increases moderately from month to month as weight increases. When the average price shifts from one weight group to another, as in the example above, only a small increase occurs.

In the third column of Table 10.2 nominal biomass value is given, while the fourth column shows the discounted biomass value (using a monthly interest rate of 0.65%).[54] As harvesting is assumed to take place at the end of the month, and the associated biomass weight in Table 10.1 is at the start of the month, we will use the biomass weight for the following month. If, for example, August of year 1 is considered, the fish have reached an average weight of 6.35 kg (from the September row of Table 10.1). The average price per kilogram is NOK19.30. Total biomass weight equals 2393 tonnes. However, as the fish lose 1% of their weight due to starving before harvest, the potential harvest weight is 2369 tonnes. Hence potential sales revenue or biomass value is NOK45.73 million. Discounted back to time zero, at 0.65% per month, the present value is NOK41.5 million.

From Table 10.2 it is noted that the nominal biomass value reaches a maximum in September of year 1, i.e. the month before the maximum biomass weight. The reason for this is that it is the biomass quantity at the start of October that is harvested in September.[55] The discounted biomass value is somewhat lower than the nominal value, and will tend to peak earlier.

---

[53]   The monthly prices have been calculated on the assumption that the size of the fish is normally distributed with the mean represented by the weights in Table 10.1. Further, based on observations in Norway, it is assumed that the standard deviation is equal to 0.2 of the mean. The price per kilogram has been set to NOK17.1 for fish less than 1 kg in weight, NOK17.8 for fish in the 1–2 kg range, NOK18.6 for the 2–3 kg range, NOK19.0 for the 3–4 kg range, NOK19.2 for the 4–5 kg range, NOK19.3 for the 5–6 kg range, NOK19.4 for the 6–7 kg range, and NOK21.8 for fish weighing more than 7 kg.

[54]   A monthly interest rate of 0.65% is equivalent to an annual interest rate of 8%.

[55]   In Chapter 9 and the Appendix the relationship between the times of maximum biomass value, biomass weight and individual weight is discussed.

10.4   Feeding costs

As feed costs are incurred every month (period), total feed cost will be the sum of the feed expenditures in every month discounted back to the release time. In this example, the following assumptions are made:

- Feed conversion ratio: 1.10 kg feed per kilogram of weight increase.[56]
- Feed price: NOK7.00/kg.

The following relationships exist:

(1) Feed consumption per month = weight increase × feed conversion ratio. This can also be expressed in symbols as:

$$F_t N_t = \left(w_{t+1} - w_t \right) f N_t$$

where F_t is feed consumption per fish per month, N_t as before the number of fish, $(w_{t+1} - w_t)$ is the weight increase per month and f is the feed conversion ratio, which is assumed to be constant. It can be seen that the quantity of feed depends on the weight increase, the feed conversion ratio and the number of fish.

(2) Feed costs per month (nominal) = feed consumption × feed price. This can also be expressed as $C_f F_t N_t$, where C_f is the feed price per kilogram. By summing feed costs over the lifetime of the fish and discounting back to the time of release, the following expression is derived:

$$\text{Present value of feed costs}: \sum_{t=0}^{n} \frac{C_f F_t N_t}{\left(1 + r \right)^t}$$

The calculations are illustrated in Table 10.3.

In order to find feed quantity per month, fish growth must be evaluated. Weight increase per fish is given in the third column of Table 10.1. With regard to the number of fish, mortality must be taken into account, as some fish will die every month. When the fish actually die is then important. The assumption here is that this occurs at the end of the month. Fish will therefore be fed for the whole month in which they die. In reality this implies a slight overestimation of feeding costs.

In May of year 0, 500 000 smolts are released with an individual weight of 106 g. In June the weight has increased to 171 g, so that the weight increase

---

[56]   At times this is separated between the economic feed conversion ratio, which is the value of the feed put into the pen, and the biological feed conversion ratio, which is what is consumed by the fish. We use the economic feed conversion ratio, as this is what is relevant for the farmer.

Table 10.3   Feed quantity (kg) and costs (thousand NOK).

Month	Feed quantity (kg) F_tN_t	Feed cost per month (nominal) $C_tF_tN_t$	Feed cost per month (discounted) $\dfrac{C_tF_tN_t}{(1+r)^t}$	Total feed costs (discounted) $\sum\limits_{t=0}^{n} \dfrac{C_fF_tN_t}{(1+r)^t}$
Year 0				
May	35 585	249.1	249.1	249.1
June	67 689	473.8	470.8	719.9
July	124 431	871.0	859.8	1 579.7
August	168 908	1 182.4	1 159.6	2 739.3
September	198 247	1 387.7	1 352.2	4 091.5
October	196 725	1 377.1	1 333.2	5 424.7
November	194 931	1 364.5	1 312.5	6 737.2
December	191 461	1 340.2	1 280.8	8 018.0
Year 1				
January	161 276	1 128.9	1 071.9	9 089.9
February	152 143	1 065.0	1 004.7	10 094.5
March	148 872	1 042.1	976.7	11 071.3
April	191 193	1 338.4	1 246.3	12 317.6
May	235 613	1 649.3	1 525.9	13 843.5
June	308 030	2 156.2	1 982.0	15 825.5
July	365 932	2 561.5	2 339.4	18 164.9
August	386 628	2 706.4	2 455.8	20 620.7
September	376 193	2 633.3	2 374.0	22 994.7
October	329 510	2 306.6	2 066.0	25 060.7
November	289 462	2 026.2	1 803.2	26 863.9
December	248 211	1 737.5	1 536.2	28 400.1

is 65 g.[57] With a 1.1 conversion ratio, 71.2 g of feed per fish are required. All 500 000 fish are actually fed in May, so the total feed quantity equals 35 585 tonnes. At NOK7/kg, this generates a feed cost for May of NOK249 096. As May is defined as time zero, there is no discounting for this month.

On the other hand, if January of year 1 is taken as the point of reference, the weight increase per fish during this month is 0.307 kg. A feed factor of 1.1 implies a feed quantity of 0.337 kg per fish. At the beginning of January, there are 476 967 fish in the pens. All are fed all month, including the ones that die. Total feed quantity then is 161 276 tonnes and, at NOK7/kg, feed

---

[57]   There is a rounding error in 65 g. The exact number is 0.647, which is necessary if one is to replicate the table exactly.

Table 10.4   Present value of cash flows from harvesting in January to December of year 1 (million NOK).

Month	Discounted sales revenue	Discounted harvesting costs	Discounted feed costs	Present value (earnings)
Year 1				
January	21.4	6.7	9.1	5.6
February	23.4	6.6	10.1	6.8
March	25.6	6.4	11.1	8.1
April	28.0	6.2	12.3	9.4
May	31.0	6.1	13.8	11.1
June	34.6	5.8	15.8	13.0
July	38.5	5.5	18.2	14.8
August	41.5	5.1	20.6	15.7
September	43.1	4.7	23.0	15.5
October	42.5	4.1	25.1	13.3
November	40.9	3.6	26.9	10.5
December	38.5	3.1	28.4	7.0

costs are NOK1 128 900. Discounting back to the time of release results in a present value of NOK1 071 900.

The last column of Table 10.3 shows the sum of the feed costs discounted back to the time of release. In January of year 1 total feed costs (discounted) are NOK9 089 900.

10.5   Net present value

As already shown, there are in theory 20 possible investment alternatives: harvesting (or selling the fish) from month zero through month 19. However, as noted above harvesting is relevant only in a shorter period. This analysis will evaluate harvesting in the period January to December of year 1, i.e. 12 alternatives. The present values of cash flows from harvesting in those 12 months will be compared to find the alternative that maximises the present value of the investment.

For this purpose Table 10.4 is used as it shows the present value of cash flows (revenues and costs) from harvesting in the respective 12 months. The discounted sales revenue (biomass value) column gives potential sales revenues by month of harvest. It reaches a maximum of NOK43.1 million in September of year 1. The value is taken from Table 10.2.

The assumption is made that harvesting costs NOK15.00 per fish. It is assumed that harvesting costs depend on the number of fish harvested and not the biomass weight of the fish. Harvesting costs are of relevance only if harvesting actually takes place. If harvesting occurs at the end of January of year 1, there are 470 290 fish (see Table 10.1, the row for February). With a

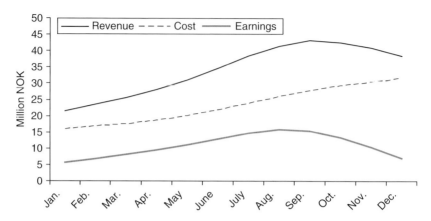

Figure 10.2   Discounted value, costs and the present value of cash flows (earnings).

cost of NOK15 per fish, nominal harvesting costs are NOK7 054 350. Discounted back to the time of release, it is NOK6.7 million. Note that harvesting costs diminish over time as the number of fish is reduced due to natural mortality. Moreover, the present value diminishes due to discounting. By postponing harvesting, costs are reduced accordingly. The discounted feed costs are shown in the fourth column. They are taken from Table 10.3 and increase over time.

The present value of harvesting in the respective months is shown in the final column of Table 10.4. The table shows that present value is maximised when harvesting at the end of August of year 1. This is 2 months before the month of maximum biomass weight (Table 10.1) and 1 month before the maximum discounted biomass value. This is also presented graphically in Figure 10.2, where the present value of the investment (earnings) is shown along with total costs (harvesting plus feed costs) and discounted biomass value. The reason the present value of the investment reaches its maximum in August is that costs increase so much from September to October that net earnings in this and later months become negative even though the biomass increases.

The present value in August of year 1 is NOK15.7 million. If the fish were harvested 1 month earlier, it would be NOK0.9 million lower, whereas harvesting in September would imply a loss of NOK0.2 million. This is a loss of 0.1–0.6% of the maximum present value, which is not very large. Hence, the present value close to the optimum is not very sensitive to exact timing. This can also be seen in Figure 10.2, where the present value (earnings) curve is shown to be rather flat, and much flatter than the revenue curve. This is caused by the steadily increasing cost curve.

A change in parameter values might imply a change in the harvesting plan. However, due to the considerable calculations involved, no sensitivity analysis was undertaken. The directions of change will be as discussed in Chapter 9.

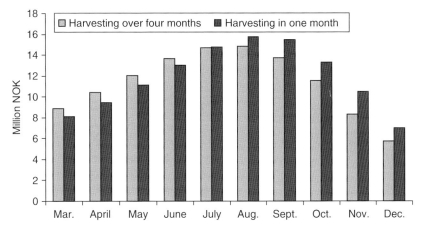

Figure 10.3 The present value of cash flows of harvesting in 1 month and harvesting in 4 months for the period March year 1 to November year 2.

10.6   Selective harvesting

With the assumptions underlying this analysis, it is optimal for the fish farmer to harvest all fish in the same month. With other assumptions it might be optimal to spread the harvesting over time (selective harvesting). This could be due to differences in growth (see Chapter 9), seasonal price variations, or a desire to spread risk. Moreover, supply may have less impact on price when spread over a longer time. Lower harvesting costs can be envisioned if harvesting is spread over time, as this work could then be undertaken to a larger degree by the normal labour force. This will typically be a more important issue in Scotland, where on-site harvesting is common, while it is not in Chile and Norway.

In Figure 10.3, the present values of cash flows from harvesting in 1 month and over 4 months are shown. The 1-month value is compared with the 4-month value by using the third month of any given 4-month span as the comparison point. The 1-month alternative is the same as the one in Table 10.4. In the second alternative, about 25% of the biomass weight is harvested per month over 4 months.[58] The price and cost parameters are the same.

The figure illustrates that, overall, harvesting over 4 months gives a slightly lower maximum present value than harvesting all the fish in 1 month. The period of maximum present value under selective harvesting is July to October of year 1 with NOK14.8 million, compared with September

---

[58]   In the first month 25% of the fish are harvested, in the second month 33% of the remaining fish are harvested, then 50% of the remaining fish in the third month and the remainder of the fish in the final month. As before, harvesting is assumed to occur at the end of the month, so that fish are fed up to the month before harvesting.

of year 1 and a value of NOK15.7 million when all fish are harvested in the same month. Interestingly, the difference between these two alternatives of about 5% is relatively small. This result is not surprising, as minor deviations from optimal harvesting time were shown in Figure 10.2 to imply only a small reduction in present value. The practical consequence of this is that harvesting spread over a somewhat longer period can be an alternative to harvesting in 1 month. This strategy can reduce risk when prices and growth are difficult to predict.

There are several reasons why selective harvesting can be optimal. As mentioned in Chapter 9, this can be the case when fish growth differs. Another biological reason for early harvesting would be premature sexual maturation. Earlier harvesting certainly reduces the risk. However, such a policy is questionable in practice as it is not likely that harvesting costs will remain constant. It is also infeasible to stop feeding some fish in one pen because they are to be harvested, and harvesting will accordingly take place on a pen-by-pen basis. So-called high-cropping, or just harvesting the largest fish from a pen, is therefore not a common practice.

Spreading the harvest over a longer period might enable the farmer to undertake all harvesting with the normal labour force, whereas hiring additional labour is required when harvesting in a short period. As such, reducing costs by spreading harvesting over time may by itself make this an optimal policy if the farm in question has the facilities available. Even when slaughtering is outsourced to a specialised plant, it may be optimal to spread out harvesting, as the capacity of such plants is limited.

11 Investment in a Salmon Farm

While the biomass growth and harvesting decision for a single release of fish as presented in Chapter 10 are at the core of the operation of any fish farm, it is still only one of several important parts. This chapter considers the other main parts: the investments in the physical facilities of a salmon farm, the labour and management, the purchase of smolts and the working capital necessary to operate the firm. The results from Chapter 10 are taken as given when it comes to operating costs and production, and are used as inputs in the analysis. The analysis is conducted on an annual basis, because this is appropriate when considering long-term decisions. It also makes the analysis more tractable. The analysis in this chapter will be simplified in various ways. The purpose is to illustrate the main principles of an analysis of this kind, not to be accurate on every detail.

In section 11.1 we will set up a production plan for one salmon aquaculture firm with the capacity to release 1 million smolts each year, and also consider the variable costs that were not taken into account (or were treated as fixed) in Chapter 10. This includes the costs of labour, management and smolts, as well as an overview over the firm's investment costs. In section 11.2 we consider the capital requirements for the company with respect to both financing the plant and working capital. On this basis, we establish a liquidity budget for the first 2 years of operation, before in section 11.3 we present an estimate of the cost of production per kilogram of salmon. Section 11.4 discusses the investment decisions of an investor who would like to start up a salmon farm. Section 11.5 deals with licence values. In section 11.6 the valuation of an operating salmon farming firm is discussed, and thereby the decisions of any investor considering the purchase of such a company.

11.1 A production plan

We start by developing a plan for the new firm's production of salmon. Assume that the firm has two plants, A and B. This is now standard practice in most salmon-producing companies to allow harvest in all years. Further

The Economics of Salmon Aquaculture, Second Edition. Frank Asche and Trond Bjørndal.
© 2011 Frank Asche and Trond Bjørndal. Published 2011 by Blackwell Publishing Ltd.

assume that each plant has a capacity of 1 million smolts a year, that all necessary permits and licences have been obtained, and that an environmental impact analysis has been undertaken. To smooth the harvest over the year, the smolts in each plant will be released at two different times, one batch in May and another batch in October. Also assume that the company has access to three sites, so that one site will lay fallow for a year every third year.

We can summarise the main assumptions as follows:

Dates and periods
- Start up date for Plant A is assumed to be 1 May of year 0 when 500 000 smolts are released. A second batch, also 500 000 smolts, is released on 1 October. For Plant A, there will be no smolt release in year 1; in year 2, batches will again be released in May and October.
- Plant B starts operations in year 1, releasing batches of 500 000 smolts in May and October. Plant B will have no release in year 2; its next release will be in year 3.
- Each batch of smolts matures after 16 months, so the May batch is harvested at the end of August the following year, while the October batch is harvested at the end of January two calendar years later. All batches of smolts are the same in number as the first batch, and with identical growth patterns.

Weight and growth
- Growth data are as in section 10.2, and it is assumed that the October batch has the same characteristics when it comes to growth as the May batch, so that the only difference is that it is released and harvested 5 months later.

Production
- The number of fish from one batch after 16 months in the sea is 376 842 with biomass weight of 2393 tonnes. As the fish are starved prior to harvesting, harvest weight is 99% of this or 2369 tonnes (cf. section 10.3, Table 10.1).

Variable costs
- Smolt price: NOK5.80 per piece.
- Labour: five farmhands (2.5 at each plant) at NOK600 000 per man-year including social costs.
- Maintenance and miscellaneous costs: NOK1 500 000 per year.

Fixed cost
- Management: NOK1 000 000 per year. This includes one manager, associated management and office costs.

The production plan is given in Table 11.1 and is illustrated in Figure 11.1, which gives a timeline for the first 48 months of operations, showing the

Table 11.1   Production plan for salmon farm.

| Year | Harvest from yearclass | | | | | | | | Harvest per year |
	Yearclass 1 May	Yearclass 1 September	Yearclass 2 May	Yearclass 2 September	Yearclass 3 May	Yearclass 3 September	Yearclass 4 May	Yearclass 4 September	
0	0	0	0	0	0	0	0	0	0
1	2369	0	0	0	0	0	0	0	2369
2	0	2369	2369	0	0	0	0	0	4738
3	0	0	0	2369	2369	0	0	0	4738
4	0	0	0	0	0	2369	2369	0	4738
5	2369	0	0	0	0	0	0	2369	4738
6	0	2369	2369	0	0	0	0	0	4738
	2369	2369	2369	2369	2369	2369	2369	2369	

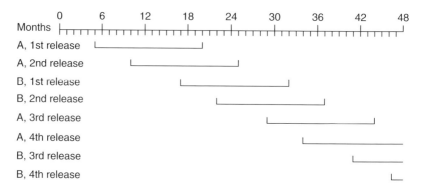

Figure 11.1 Production plan years 0–4. A, Plant A; B, Plant B.

production cycles for different releases of smolts. As noted, Plant A releases smolts in May and October in year 0. The first batch is harvested at the end of August year 1, at a biomass weight of 2369 tonnes, while the second batch is harvested in early year 2. Plant B starts operations in year 1. The May yearclass is harvested at the end of August year 2, while the October yearclass is harvested in early year 3. Thus year 2 is the first year with full production of 4738 tonnes. Table 11.1 shows production until year six but operations are, of course, still ongoing.

The main components of the physical structures and their cost for each plant (handling 1 million smolts) can be described as follows:

- Eight pens (120 m circumference) with nets and moorings: NOK5 200 000.
- Feed barge with operating facilities and storage capacity for 320 tonnes of feed: NOK6 000 000.
- Automated feeding system: NOK1 400 000.
- Camera and environmental sensor systems: NOK1 000 000.
- Other equipment and expenses: NOK2 000 000.
- Total costs: NOK15 600 000.

Two releases of smolts each year require four pens for each release. Other equipment and expenses include light systems, anchoring the barge, operational software, etc. Two plants require a total capital investment of NOK31.2 million, with equal amounts in year 0 and year 1.

The investors must also cover the cost of acquiring the necessary permits. This cost will differ considerably between countries and regions, but may amount to a substantial sum. The cost of environmental impact assessments, legal costs, and other start-up costs are set at NOK5 million.

If a farming licence must be bought, this is an additional cost. In Norway the cost of one licence is NOK8 million (Chapter 3); as the firm requires four licences, the cost is NOK32 million. In Scotland and Chile, where a farm has to pay an annual fee, this cost will add to the annual operating expenses.

On this basis, we can establish the following investment costs:

Year 0
- Environmental assessment, legal costs, etc.: NOK5.0 million.
- Licences: NOK32.0 million.
- Facility investments: NOK15.6 million.

Year 1
- Facility investments: NOK15.6 million.

This means that total investments will amount to NOK68.2 million.

11.2   A liquidity budget

As explained above, substantial amounts will be invested in physical plants, licences, etc. in years 0 and 1. Production commences with the release of the first batch of smolts in May, year 0, but the fish are not harvested until the autumn of year 1. Full production is not achieved until year 2. Over this period the investors must be able to fund not only investments but also operations. Unless they are able to do so, an otherwise profitable investment cannot be undertaken. This is why the liquidity situation of the firm is crucial, in particular during the first years of operation. The fact that most of the aquaculture operation's expenses occur before any revenue is generated means that the company not only needs to finance investments, but also the company's working capital. In this section, a liquidity budget will be derived for the newly developed fish farm to show the capital requirements.

According to the production plan, salmon are harvested 16 months after smolts are released. Throughout this period smolts, feed, labour and other costs must be paid for by the firm. These are issues relating to the **working capital** of the firm. The working capital can be obtained in three different ways, and often it will be a combination. First, the company's owners may provide part of the working capital as part of the firm's equity. Second, the firm can get credit from an external source such as a bank, or its suppliers. Finally, the firm can retain operating revenues. For a newly developed farm, the third option is not available. However, for an aquaculture firm that has been in operation for some time, part of the sales revenues will normally be retained to cover operating costs.

In order to compute the liquidity budget, information about costs and revenues is used. In particular, the start-up and recurring costs and the sales revenue, based on the production plan, are involved in forming the liquidity budget.

We assume that the farm in question is a turnkey facility. Year 0 investments of NOK52.6 million are made on 1 May, while year 1 investments of NOK15.6 million are made on 1 May. The following assumptions are made regarding financing:

- Facility investments are financed by equity (50%) and by a long-term bank loan (50%). Thus, the first bank loan of NOK26.3 million is taken up on 1 May of year 0, with a second bank loan of NOK7.8 million on 1 May of year 1.
- The loan is serviced twice a year, at the end of June and December.
- For each bank loan assume no repayments on the principal for the first 2 years; thereafter repayment over 10 years. Repayment is assumed to be by the linear method. Thus, for the first loan, a repayment of NOK1.315 million is made every half year, starting at the end of June of year 2. For the second loan, a repayment of NOK0.365 million is made every half year, starting at the end of June of year 3.
- The bank interest rate as well as the required return to investors is 8% per annum.

For labour and management, make the following assumptions:

- The manager is employed 1 May of year 0.
- Three workers are employed on 1 May of year 0, the remaining two on 1 May of year 1.

Maintenance and miscellaneous costs are incurred on a linear basis over time, starting 1 May of year 0.

Let us also repeat the most important assumptions from Chapter 10:

- Average sales price: NOK19.30/kg.
- Harvest quantity: 2369 tonnes in year 1; 4738 tonnes per year in year 2 and subsequent years.
- Feed conversion ratio: 1.10 kg feed per kilogram of weight increase.
- Feed price: NOK7.00/kg.
- Harvesting cost: NOK15.00 per fish.

All costs are paid in the period they are incurred in, while sales revenue is received in the period the fish are sold.

In the first 16 months, operating costs are incurred while sales revenue is zero. As there are no sales during this period, the farm runs at an operating loss. The operating costs are financed by a line of credit. The interest rate applied is 8% per annum. The interest charges are calculated on the basis of average utilisation of the line of credit.

Table 11.2 shows the liquidity budget of the farm for the first 2 years, years 0 and 1; however, the budget is established on a half-yearly basis. The first line in Table 11.2 shows the sales revenue. The first sales revenue of

Table 11.2   Liquidity budget (million NOK).

	First half-year	Second half-year	Third half-year	Fourth half-year
Sales revenue	–	–	–	45.700
Smolts	–2.900	–2.900	–2.900	–2.900
Feed	–0.723	–9.117	–16.884	–24.187
Labour, maintenance and management	–0.717	–2.150	–2.250	–2.750
Harvesting costs	–	–	–	–5.653
Financial payments	–0.351	–1.052	–1.156	–1.364
Capital requirement current period prior to interest charges	–4.691	–15.219	–23.190	8.846
Capital requirements previous periods	–	–4.739	–19.958	–43.148
Capital requirement prior to interest charges	–4.691	–19.958	–43.148	–34.302
Interest charge, line of credit	–48	–333	–1.220	–1.359
Capital requirement	–4.739	–20.291	–44.368	–35.661
Change in liquidity	–4.739	–15.552	–24.077	8.707

NOK45.7 million is obtained in the fourth half-year. This is from the sale of the first batch of smolts released in Plant A in May, year 0, and sold at the end of August in year 1. From year 2 onwards a steady-state annual income of NOK91.4 million will be obtained.

The second line shows smolt purchases, which are incurred from start-up of the farm. Line three shows feed costs. They are a reflection of the release of smolts, the first batches in May and October year 0, followed by two more batches in year 1. In the fourth half-year, the farm is feeding a total of four batches. This means that feed costs increase over time. Steady-state annual feed costs of NOK43.8 million will be reached in year 2.

While management, maintenance and other costs are incurred from day one, only three workers are employed in year 0 and an additional two in year 1. Therefore, these costs (line four) also increase over the first three periods shown in the table.

Harvesting costs are incurred for the first time in the third half-year, when 376 842 fish of the first batch are harvested (line five; cf. Table 10.1) at a cost of NOK5.635 million.

Financial payments are 'limited' in the first four half-year periods (line six), as the firm pays only interest on the bank loans. However, in year 2, the firm starts paying off the first loan with a total of NOK2.63 million, so that total financial payments amount to NOK5.3 million. In year 3, repayments of the second loan also start with NOK0.73 million, so that financial payments increase to NOK5.8 million. However, this amount will decline in subsequent years because of the linear method of repaying the loans.

Sales revenue minus operating costs (smolts, feed, labour, maintenance, management and harvesting costs) and financial payments equals capital requirements of the current period, prior to interest charges (line seven). In the first half-year, the capital requirement is NOK4.691 million, increasing to NOK23.190 million in the third half-year. A positive cash flow of NOK8.846 million is generated for the first time in the fourth half-year, when the first batch of salmon is harvested and sold.

Capital requirement of the previous period is given in line eight. Capital requirements of the previous periods must be added to that of the current period to arrive at accumulated capital requirement (line nine).

Interest charge on the line of credit is given in line 10. For smolts, this is calculated from the date they are actually purchased. Feed, labour, maintenance and management costs are assumed to be incurred on a linear basis. As everything is financed from a line of credit, capital requirement of the previous period must also be considered when estimating the interest charge.

The interest charge on the line of credit increases from NOK48 000 in the first half-year to NOK1.359 million in the fourth.

'Capital requirement' (penultimate line) gives the combined capital requirement to finance the cash flow incurred by both investments and operations. It is increasing over time from NOK4.739 million at the end of the first half-year to NOK44.368 million at the end of the third; it is negative even after the end of the fourth half-year period. Under the assumptions made, 'capital requirement' will turn positive in year 2, which means the firm will not only be able to pay off the line of credit, but also generate a net positive liquidity from this period onwards.

The last line shows the change in liquidity. This is equal to the change in capital requirement (penultinate line) from one period to the next.

The first harvest takes place at the end of August in year 1, i.e. in the fourth half-year period. At this point the capital requirement exceeds NOK50 million. Total investment costs were shown to amount to NOK68.2 million (section 11.2). Under the assumptions here, the total capital requirements for facility investments and operations are in excess of NOK105 million, at its maximum. Thus, only considering facility investments gives a very misleading picture of maximum capital requirements for a newly developed firm.

When establishing the liquidity budget, it is assumed that sales revenue is used to pay down the line of credit.

In the case under consideration, production proceeds without problems of any kind. Aquaculture is often a risky business. An unexpected disease outbreak can wipe out an entire batch of fish and cause great financial loss if the insurance coverage is limited. Less serious incidents can still be very costly. Over the years, many firms have gone into bankruptcy for this very reason, as they have not had the required funds to stave off this kind of financial emergency, even if the firm is profitable in the long run. This highlights the importance of liquidity planning. For practical purposes, the analysis presented here would need to be extended to consider the consequences of, for example, higher mortality and lower growth than what had been planned for, and a monthly or even weekly model would need to be made use of. The consequences of an entire batch being wiped out should also be analysed.

Harvesting costs and smolt costs are incurred only twice a year. Costs related to factors such as labour, management and maintenance are incurred in all time periods. This is partly due to the fact that it would be difficult to maintain high-quality employees if they are laid off after harvest in all production cycles. In addition, management and maintenance functions do not disappear even if there are no fish in the pens, and there are a number of specific tasks that need to be performed when there are no fish in the pens, for example major net and pen maintenance.

11.3   Cost of production

Based on the information presented, we can derive this company's production cost per kilogram. This cost of production analysis assumes the firm is in steady state, i.e. in full production. This is appropriate, as it will indicate the kind of price the firm will need to fetch for its products in order to be financially viable over time. From section 11.1, the steady-state production of this company is 4738 tonnes per year. We also have the information we need for operating costs.

To find the average cost of production, we must also find capital costs, which consist of depreciation and interest on invested capital. In section 11.1, it was shown that total investments amount to NOK68.2 million. The physical structures depreciate over time. This can be dealt with in different ways. One can estimate annual depreciation according to different methods. The different parts of the physical structure have lifespans ranging from 8 years for the automated feeding system to 25 years for the feed barge (Table 11.3). Often depreciation is based on a linear method or by a certain percentage of remaining value per year, which means that the annual depreciation will vary over time. Alternatively, one can estimate an annualised depreciation and capital cost. This makes the use of capital, in terms of depreciation and the alternative cost of capital, into a steady annualised cost. This is done in Table 11.3 with annual depreciation and interest (8% per annum) on the facility investments. Licenses do not depreciate and for them only interest is charged. The total is then almost NOK7.2 million per year. These estimations utilise the annuity method, implying that the sum of depreciation and interest is constant per year.[59] Alternatively, this can be considered the average amount of depreciation and interest over time.

---

[59]   House mortgages are commonly repaid according to the annuity method. This means that the monthly amount paid for interest and repayment on capital is constant over time (unless the interest rate changes).

Table 11.3   Annual depreciation and interest on facility investments.

Item	Repayment time (years)	Cost (NOK)	Depreciation and interest (NOK)
Environmental assessment and legal costs	–	5 000 000	400 000
Licences	–	32 000 000	2 560 000
Pens	10	10 400 000	1 549 902
Feed barge	25	12 000 000	1 124 142
Automated feeding system	8	2 800 000	487 244
Camera and environmental sensor systems	8	2 000 000	348 032
Other equipment and expenses	8	4 000 000	696 064
Total		68 200 000	7 165 384

Table 11.4   Annual cost of production for 4738 tonnes and average cost per kg salmon.

Cost category	Total cost (thousand NOK)	Average cost per kg (NOK/kg)	Percentage of total
Smolts	5 800	1.22	7.5%
Feed costs	43 787	9.24	56.8%
Labour	3 000	0.63	3.9%
Maintenance	1 500	0.32	2.0%
Harvesting costs	11 305	2.39	14.7%
Management	1 000	0.21	1.3%
Interest on working capital	3 506	0.74	4.6%
Total variable costs	69 898	14.75	90.7%
Total capital cost	7 165	1.51	9.3%
Total	77 063	16.26	100%

Once steady-state production has been achieved, total annual cost of production is NOK77.063 million (Table 11.4). This is composed of variable costs, including interest on working capital, at NOK69.898 million, representing 90.7% of the total, and capital costs at NOK7.165 million, representing 9.3% of the total. With a production of 4738 tonnes, average cost of production is NOK16.26/kg.

The variable costs include all the costs that are related to producing salmon given that the plants are available. They are broken down to the main categories that have been discussed earlier: smolt, feed, labour, maintenance, harvest and management costs as well as the cost of the working capital.

Interest on working capital (variable and management costs, but excluding harvesting costs) is estimated on the basis of a 16-month production cycle. Interest on smolt cost is estimated for 16 months. As a simplification, all other costs are incurred in a linear fashion over the production cycle.

Variable cost per kilogram of salmon produced is NOK14.75. Feed, with a cost share of 56.8%, is clearly the largest factor. Harvesting cost, with a 14.7% cost share, is second. In is interesting to note that interest on working capital is also quite substantial, with a 4.6% cost share.

The capital cost is the depreciation and return or interest on the plant itself. These costs are based on Table 11.3, and contain the categories permits, licences, pens, feed barge and other equipment. Relative to operating costs, the capital costs are small at NOK1.51/kg, making up 9.3% of cost of production.

The figures presented here are somewhat different from the actual cost of production in Norwegian salmon aquaculture (as shown in Table 4.1). There are two main reasons for this. First, in our figures harvesting cost is included, while they are not in the official figures reported in Chapter 4. Second, Table 4.1 gives cost of production figures per year, where production is defined as sales plus inventory changes for the stock of fish, while in this chapter we use numbers that are averages for the 16-month production cycle.

11.4   Investing in a new aquaculture company

For potential investors in this salmon firm, what is of interest is whether they will achieve an acceptable return on the capital they invest. An investment will only be undertaken if the investor achieves that kind of return on money invested. This is the issue that will be investigated in this section.

In Chapter 10, we evaluated the present value of cash flows generated by an investment in a yearclass of fish to determine the optimal time of harvesting. The approach is the same here, but also includes cash flows related to the factors that were excluded from the analysis in Chapter 10.

Evaluating the value of the company requires information about the cash flow it will generate. We can identify three types of cash flows, namely those related to (i) investments and reinvestments, (ii) revenues from the sale of fish and (iii) operating costs such as feed and labour. Revenues are received in the same period as the fish are sold, while feed, labour and other inputs need to be paid in the period they are incurred. The objective for the investor is then to determine if the initial investment will provide a positive expected present value (EPV) of cash flows given the investor's required rate of return (interest rate). This amounts to:

$$EPV = \sum_{t=0}^{t=n} \frac{-I_t}{(1+r)^t} + \sum_{t=0}^{t=n} \frac{A_t}{(1+r)^t} = \sum_{t=0}^{t=n} \frac{-I_t}{(1+r)^t} + \sum_{t=0}^{t=n} \frac{V_t - TC_t}{(1+r)^t}$$

Table 11.5   Investments per year.

Year	Amount (million NOK)	Purpose
0	52.6	Plan A and licences
1	15.6	Plant B
8	4.4	Reinvestment
9	4.4	Reinvestment
10	5.2	Reinvestment pens
11	5.2	Reinvestment pens
25	6.0	Feed barge
26	6.0	Feed barge

Investments and reinvestments are undertaken in several periods, with I_t being investment in period t. $A_t = V_t - TC_t$ is net cash flow in period t, V_t is revenues in period t and TC_t is cost in period t, with r as the periodic rate of return. For this type of analysis, the time span for each period will normally be a year, and that is the unit used here. In this formula, TC_t represents operating costs. Depreciation is not included, as it involves no cash flows. The interest represents the investor's required return on capital. This will be explained in more detail below.

To carry out this analysis, the information in sections 11.1, 11.2 and 11.3 will be used as building blocks to determine the cash flows arising as a consequence of the investment. In year 0, there are investments in licences amounting to NOK37 million, while the facilities for Plant A cost NOK15.6 million. In year 1, investments in Plant B amount to NOK15.6 million. Thus in the formula above I_0 is NOK52.6 million, while I_1 is NOK15.6 million.

As noted above, the different parts of the facility will have different lifespans. The automated feeding system, camera and environmental sensor systems, as well as other equipment and expenses, all have a depreciation period of 8 years (Table 11.2). For simplicity, assume the production technology remains unchanged over time. This implies that, for Plant A, these systems will be reinvested in year 8 at a cost of NOK4.4 million, while for Plant B reinvestment will be in year 9. In years 10 and 11 the pens will need to be replaced, while new barges are required in years 25 and 26. Initial investments and reinvestments in years 0–26 are shown in Table 11.5. It is assumed that the scrap value of fixed assets is zero when they come to the end of their depreciation periods. A positive scrap value would count as a sales revenue.

There is a need to assess revenues and costs over time. In the formula above, V_t is revenues in period t and TC_t is cost in period t. V_t is sales revenue and is found in the cash flow budget in Table 11.3: NOK45.7 million in year 1, increasing to NOK91.4 million on an annual basis as of year 2.

Table 11.6   Annual cash flows: revenues, expenses and investments.

Year	Revenues (million NOK)	Expenses (million NOK)	Investments (million NOK)
0	0	18.507	40.6
1	45.7	57.524	15.6
2	91.4	66.379	–
3	91.4	66.379	–
4	91.4	66.379	–
5	91.4	66.379	–
6	91.4	66.379	–
7	91.4	66.379	–
8	91.4	66.379	4.4
9	91.4	66.379	4.4
10	91.4	66.379	5.2
11	91.4	66.379	5.2
12: terminal value of cash flows	1142.5	829.7	27.4

Annual revenues, operating costs and investments for years 0–11 are summarised in Table 11.6.

The following operating costs have been included: smolts, feed, labour, maintenance and other costs, and harvesting costs. Financial costs (interest on fixed investments and working capital as well as depreciation) are not included, as the cost of capital is accounted for by including the investments (and reinvestments) and the discount rate, representing the required return on capital. Moreover, depreciation does not involve any cash outlay. Costs increase from NOK18.507 million in year 0 to NOK57.524 million in year 1 and a steady-state level of NOK66.379 million in year 2.

When valuing a company, after some point in time the cash flow will not change between periods as there is no new information available. It is then common to estimate the terminal value, which is the value of the cash flow from that point on to infinity. This is given as A/r, where A is the annual cash flow per year and r is the discount rate. This is provided in row 12 for total revenue as well as expenses and investments. To compute the terminal value, we assume that from year 12 onwards the firm will continue producing 4738 tonnes of salmon per year, generating an annual revenue of NOK91.4 million. Capitalised at 8%, this gives a terminal value of NOK1142.5 million.

Operating costs per kilogram equal NOK14.75 – 0.74 = NOK14.01/kg. This is equal to variable costs minus interest on working capital (cf. Table 11.4). Thus, annual operating costs will represent NOK66.379 million. Capitalised at 8%, this gives a terminal value of NOK829.7 million.

As explained above, the various components of the physical investments need to be replaced over time through reinvestments. This will be the case also after year 12. To take care of this, the present value, as of year 12, of all future investments has been estimated.[60] This amounts to NOK27.4 million.

The terminal value in year 12 of the cash flows of the firm then becomes NOK1142.5 – 829.7 – 27.4 = NOK285.4 million. This represents the value of the firm in year 12.

All values are discounted to year 0 according to the formula above. Year 0 values are not discounted; year 1 values are discounted 1 year, etc. In doing so, it is assumed that all revenues and expenses are incurred on a linear basis throughout the year. This is not quite correct, as smolt and harvesting costs are incurred twice a year, as are sales revenues. Regardless, this simplification will have little effect on the result.

Various industries exhibit different degrees of risk. In salmon aquaculture, one can identify production risk as well as risk about the level of prices in the future. In an investment analysis such as this, it is therefore often common to include a risk premium as part of the required rate of return. Nevertheless, as a simplification we will, as before, assume that the annual interest rate is 8%. This is the approach that would be taken by a well-diversified investor. Thus, this rate is used both for loans and as the required return on investment, and would be the rate used by a well-diversified investor.

The net present value of all cash flows associated with the investment is NOK175.1 million. This is composed of the net present value of revenues, expenses and investments in years 0–11 as presented in Table 11.6 (NOK61.8 million) and the present value of the terminal values of operations and investments (NOK113.3 million). This indicates a very profitable investment. However, one should be aware that this is a company that runs smoothly without any problems, and there is no cyclicity in the cash flow. As shown in previous chapters, both these assumptions are clearly unreasonable. Still, as salmon production has been growing rapidly for more than two decades, on average the industry must have been very profitable.

A number of simplifying assumptions have been made in the estimation of the present value. It has been assumed the price and cost of production remain constant over time, while it was shown in Chapter 4 how these two variables have been declining over time. There is no reason to expect that this process will not continue, albeit at a slower pace than hitherto. The technology has also been assumed to be unchanged, which is also unrealistic.

It would be possible to address this issue by specifying time trends for price and cost of production. However, this would not change the principles of the analysis, so it will not be done here. What can be said is that the investment is very profitable. This is likely to cause an increase in

---

[60] Reinvestments up to year 55 have been considered, but not those after year 55. Because of discounting, the error involved is very small.

production which will have a dampening effect on profitability. Thus, if anything, the present value may be overestimated.

In the analysis presented, a company size was chosen that is common for new operations today. Larger firms tend to operate several plants similar to the one presented here. However, we do not know what the optimal size is. In principle, this question could be investigated by undertaking an investment analysis for plants of different sizes to compare their economic performance. We do not have data to allow us to carry out such an analysis of scale economies, but if it were to be done, an analysis similar to the one carried out in this section would be done for different scales. The best size for a specific location would be the one with the highest expected present value.

There are certainly areas where economies of scale are important. This will, for instance, be the case for feed storage, surveillance equipment and, often, labour and management. However, there also tend to be places where there can be diseconomies of scale. For instance, harvesting is more difficult from a large pen, and with a sufficiently large pen the construction cost increases faster than capacity because of technical limitations. Furthermore, risk also tends to increase. Finally, the size of each pen and the stocking density will influence fish growth.

Different countries have different tax systems. This must be taken into account when the investment is undertaken. However, often the effect of taxes will be different for investors depending on their tax position. Therefore, in most investment prospects, earnings before interest and taxes (EBIT) are presented.

11.5   Licence value

In most countries, a licence is required to start a salmon operation (cf. Chapter 3). This is different from most other industries. It implies that there may be investors who would like to invest in salmon farming but are unable to do so because they cannot get a licence. If the number of licences is restricted, licences will take on value if the production is profitable.

The licence value is the expected earnings for the farm that exceed the returns necessary to give the investors their required rate of return on their invested capital. In the example above, the company required four licences and the expected present value of the company was NOK175 million. As the investor would be indifferent about investing or not in the company if the expected present value were zero, this means that this company provides discounted earnings that are NOK175 million higher than what is necessary to justify the investment. This also means that the investor would be willing to pay up to NOK175 million to be able to undertake the investment. For the firm in question, four licences were required. This gives a value per licence of NOK43.75 million for the specific farm analysed.

Estimation of the licence value is complicated because the value for a specific farm can, in general, be divided into two components. The first part is the value of the right to produce salmon, and this will be identical for all production sites. The second part is the value of the site itself, as some sites are more productive than others. A site with a high growth rate will have a higher value than a site with a lower growth rate. This is similar to the situation in agriculture, where some land is more productive than other land, and therefore has a higher value. Some sites are also located closer to the market so that fish from these sites will have lower transportation costs, thus giving the farmer higher net revenues. The licence value is the value of the right to produce salmon, and is equal for all salmon producers in a region. This value is then the rent of being allowed to produce salmon. The remaining value is site specific and is at times labelled as intramarginal rents.

If licences are awarded for free, which used to be the case in Norway, the licence value will be acquired more or less as a 'gift'. As licences are and have been quite valuable, this raises issues of equity. For this reason, there have been suggestions that the government should auction licences to those interested, so that all or much of the rent would be appropriated by society at large. Alternatively, licences could be sold by the government, but this raises the question of how the price should be set. In Norway, the government now charges a fee for new licences. If this fee is less than the value of the licence, there is still going to be very high demand for licences. Moreover, those who are fortunate enough to receive a licence will still obtain a 'gift' from society at large.

While the distinction between the licence value and the intramaginal rents does not sound important, it is in practice. For instance, as discussed in Chapter 3, one must pay NOK8 million for a licence in Norway (NOK5 million in the county of Finmark). Our example company required four licences to operate at the designed quantity. The value of a licence was estimated at NOK43.75 million. Having to pay NOK8 million for the licence reduces the value of the investment. When setting the licence value, the Norwegian government is implicitly saying that they think the value of the right to operate a salmon licence is at least NOK8 million.

By setting the licence value at a lower level in the county of Finmark, the Norwegian government also says that it recognises that earnings are lower in Finmark than in the rest of the country, and therefore the value of the licence is lower there. An investor would be willing to pay an amount up to earnings exceeding a normal return, including intramarginal rent, for a licence. Those investors that have access to sites with sufficiently high intramarginal rents will be willing to pay a part of those rents to obtain a licence. The investors with access to sites with low or limited rent will, on the other hand, not find it profitable to invest in the salmon plant with the high licence fee. An excessive licence fee may therefore prevent an otherwise profitable economic activity in some parts of the country.

In recent years, Norwegian investors have been willing to pay a licence fee of NOK8 million in most of the country. This is a clear indication that, for most sites, discounted earnings are NOK8 million or higher. However, the licences in Finmark have been hard to sell, indicating that the discounted earnings are less than NOK5 million. The difference in the expected earnings in Finmark and the rest of the country can be regarded as a geographical intramarginal rent for the other regions. However, the fact that it has to be paid raises the question of why farming salmon (or increasing salmon production) is encouraged in the county of Finmark but restricted in other regions where it is more profitable. The answer is not to be found within the confines of economics, but rather within the realms of regional policy and political economy.

In Chile and Scotland, farmers pay an annual fee rather than an initial one-time fee as in Norway. While seemingly different, the two ways of collecting the fee do not have much practical relevance if the rates are set at the appropriate levels, as they will then be equivalent. The present value of a constant indefinite annual cash flow, which is the same as the terminal value discussed in the previous section, is given as net earnings/interest rate. That means that an annual fee can be set equal to a one-time licence fee, given the appropriate interest rate, as the two are equal if:

Licence fee = annual fee/interest rate

For instance, an interest rate of 8% and an annual fee of NOK640 000 are equal to a licence value of NOK8 million (640 000/0.08 = 8 000 000). Thus, the relevant variables are the magnitude of the fee and the interest rate.

11.6   Buying a fish farming company

We now turn to a somewhat different situation, and will take the position of an investor who would like to purchase an operational aquaculture firm. This investor needs to find the value of the company. To obtain this information, the investor would carry out an analysis similar to what has been done in sections 11.2–11.5. However, this investor will not have to be concerned about the initial period when the firm is not receiving any revenues, but will have to estimate the value of the company's cash flows and capital stock at the time of takeover.

With access to the company's books, the investor can then find numbers similar to those presented in Table 11.6. Thus, cash flows due to current and future harvests can be estimated and the present value calculated. The simplest estimate of the company is then obtained by assuming that these mid-cycle cash flows will go on forever. That implies that, in any year, the company will obtain revenues of NOK91.4 million and have expenses of NOK66.4 million. With a return on capital of 8%, the investor will then be willing to pay NOK312.7 million for the company, as $(91.4 - 66.4)/0.08 = 312.7$. This is the same as the terminal value of revenues minus expenses (but excluding investments) in Table 11.6.

An alternative approach can be based on Table 11.4. With an average price of NOK19.30/kg and operating costs of NOK14.01/kg, the company's profit NOK5.29/kg. With production of 4 738 000 kg the value of the company is then NOK313.3 million [$(5.29 \times 4 738 000)/0.08 = 313.3$ million]. The slight differences in the numbers provided by two approaches are due to rounding errors.

In an actual transaction, the investor is willing to pay the present value of the company, adjusted for debt and the value of the company's capital. In particular, in a due diligence process, the exact value of the fish in the pens at the time of takeover as well as the value of the facilities would have to be determined. This is important as the value of the fish varies over the year and the value of the company's production facilities will vary according to how well they have been maintained.

There are at least two additional aspects an investor would need to take into consideration. First, as seen in Chapter 4, there are substantial cycles in the industry. Furthermore, the investor will often think the profits of the firm can be improved, either by increased production or lower cost (better management or synergies).

The investor (and seller) would have the benefit of knowing the current price of salmon, and try to predict where in the cycle the industry is at the point of takeover. A cycle in the Norwegian industry is about 6 years. The investor would then try to predict the cash flow for the first 6 years before setting a terminal value. However, the investor would recognise that at the bottom of the cycle, extra funds would have to be available to make sure that the firm could cover all expenditures. Accordingly, the value of the company can vary substantially depending on whether it is purchased at the peak of a cycle relative to the trough, implying that the mid-cycle values would not be used for the first few years.

It is important to understand that to find the value of a salmon company, or a company in any cyclical industry, one must take into account where in the cycle the company is when the purchase is made. Moreover, the whole cycle must be considered to get a good estimate of the company's value. There is a tendency for investors in salmon companies listed on stock exchanges to take a shorter perspective. A company looks more attractive at the top of the cycle than at the bottom.

Normally, the analysis will be based on the company's accounts. However, an investor's willingness to pay for the firm will increase if the investor thinks that revenues can be increased or costs reduced. For instance, it is commonly believed that global population growth, together with the fact that the supply from wild fish stocks is stagnant, will lead to increased demand and thereby increased prices for farmed fish. If the investor thinks this is true, he thinks that revenues will increase and increase the cash flow from the company. Similarly, cost reductions will also increase net revenues. For instance, if the investor already owns a similar company, administration costs can be reduced because the merged companies' administration need not double. Larger companies, it is also often argued, save costs in operations such as sales and processing, because their larger production gives better capacity utilisation. While these factors are not a part of the valuation problem considered here, they are certainly relevant for some investors.

In recent years, the shares in a number of salmon companies have been listed on stock exchanges. An investor in a company's shares will analyse the company in a similar fashion as an investor who would like to purchase the whole company, as a share gives ownership to a part of the company. Hence, the share value can be determined in a similar analysis as above. This is one reason why a company's EBIT and debt are important measures when presenting information about any company.

References

Aarset, B., Asche, F. & Jensen, C.L. (2006) Simulating the impact of trade restrictions: an application to the European salmon trade. *Aquaculture Economics and Management* **10**, 201–221.

Alfnes, F., Guttormsen, A.G., Steine, G. & Kolstad, K. (2006) Consumers' willingness to pay for the color of salmon: a choice experiment with real economic incentives. *American Journal of Agricultural Economics* **88**, 1050–1061.

Andersen, T.B., Roll, K.H. & Tveterås, S. (2008) The price responsiveness of salmon supply in the short and long run. *Marine Resource Economics* **23**, 425–438.

Anderson, J.L. (1985) Market interactions between aquaculture and the common-property commercial fishery. *Marine Resource Economics* **2**, 1–24.

Anderson, J.L. (1992) Salmon market dynamics. *Marine Resource Economics* **7**, 87–88.

Anderson, J.L. (2002) Aquaculture and the future: why fisheries economists should care. *Marine Resource Economics* **17**, 133–151.

Anderson, J.L. (2003) *The International Seafood Trade*. Cambridge: Woodhead Publishing.

Anderson, J.L. & Bettancourt, S.U. (1993) A conjoint approach to model product preference: the New England market for fresh and frozen salmon. *Marine Resource Economics* **8**, 31–49.

Anderson, J.L. & Fong, Q. (1997) Aquaculture and international trade. *Aquaculture Economics and Management* **1**, 29–44.

Anderson, J.L. & Wilen, J.E. (1985) Estimating the population dynamics of coho salmon (*Oncorhynchus kisutch*) using pooled time-series and cross-sectional data. *Canadian Journal of Fisheries and Aquatic Sciences* **42**, 459–467.

Anderson, J.L. & Wilen, J.E. (1986) Implications of private salmon aquaculture on prices, production, and management of salmon resources. *American Journal of Agricultural Economics* **68**, 867–879.

Arnason, R. (1992) Optimal feeding schedules and harvesting time in aquaculture. *Marine Resource Economics* **7**, 15–35.

Arrow, K., Bolin, B., Costanza, R. *et al.* (1995) Economic growth, carrying capacity, and the environment. *Science* **268**, 520–521.

Asche, F. (1996) A system approach to the demand for salmon in the European Union. *Applied Economics* **28**, 97–101.

Asche, F. (1997) Trade disputes and productivity gains: the curse of farmed salmon production? *Marine Resource Economics* **12**, 67–73.

Asche, F. (2001) Testing the effect of an anti-dumping duty: the US salmon market. *Empirical Economics* **26**, 343–355.

Asche, F. (2008) Farming the sea. *Marine Resource Economics* **23**, 507–527.

Asche, F. & Guttormsen, A.G. (2001) Patterns in the relative price for different sizes of farmed fish. *Marine Resource Economics* **16**, 235–247.

Asche, F. & Sebulonsen, T. (1998) Salmon prices in France and the UK: does origin or market place matter? *Aquaculture Economics and Management* **2**, 21–30.

Asche, F. & Tveterås, R. (1999) Modeling production risk with a two-step procedure. *Journal of Agricultural and Resource Economics* **24**, 424–439.

Asche, F. & Tveterås, S. (2004) On the relationship between aquaculture and reduction fisheries. *Journal of Agricultural Economics* **55**, 245–265.

Asche, F. & Tveterås, S. (2008) International fish trade and exchange rates: an application to the trade with salmon and fishmeal. *Applied Economics* **40**, 1745–1755.

Asche, F., Salvanes, K.G. & Steen, F. (1997) Market delineation and demand structure. *American Journal of Agricultural Economics* **79**, 139–150.

Asche, F., Bjørndal, T. & Salvanes, K.G. (1998) The demand for salmon in the European Union: the importance of product form and origin. *Canadian Journal of Agricultural Economics* **46**, 69–82.

Asche, F., Guttormsen, A.G. & Tveterås, R. (1999a) Environmental problems, productivity and innovations in Norwegian salmon aquaculture. *Aquacultural Economics and Management* **3**, 19–30.

Asche, F., Bremnes, H. & Wessells, C.R. (1999b) Product aggregation, market integration and relationships between prices: an application to world salmon markets. *American Journal of Agricultural Economics* **81**, 568–581.

Asche, F., Guttormsen, A.G. & Tveterås, S. (2001a) Aggregation over different qualities: are there generic commodities? *Economics Bulletin* **3**, 1–6.

Asche, F., Bjørndal, T. & Young, J.A. (2001b) Market interactions for aquaculture products. *Aquaculture Economics and Management* **5**, 303–318.

Asche, F., Gordon, D.V. & Hannesson, R. (2002) Searching for price parity in the European whitefish market. *Applied Economics* **34**, 1017–1024.

Asche, F., Bjørndal, T. & Sissener, E.H. (2003) Relative productivity development in salmon aquaculture. *Marine Resource Economics* **18**, 205–210.

Asche, F., Gordon, D.V. & Hannesson, R. (2004) Tests for market integration and the law of one price: the market for whitefish in France. *Marine Resource Economics* **19**, 195–210.

Asche, F., Guttormsen, A.G., Sebulonsen, T. & Sissener, E.H. (2005) Competition between farmed and wild salmon: the Japanese salmon market. *Agricultural Economics* **33**, 333–340.

Asche, F., Kumbhakar, S. & Tveterås, R. (2007a) Testing cost versus profit functions. *Applied Economics Letters* **14**, 715–718.

Asche, F., Roll, K.H. & Tveterås, R. (2007b) Productivity growth in the supply chain: another source of competitiveness for aquaculture. *Marine Resource Economics* **22**, 329–334.

Asche, F., Jaffry, S. & Hartman, J. (2007c) Price transmission and market integration: vertical and horizontal price linkages for salmon. *Applied Economics* **39**, 2535–2545.

Asche, F., Bjørndal, T. & Gordon, D.V. (2007d) Studies in the demand structure for fish and seafood products. In: Weintraub, A., Romero, C., Bjørndal, T. & Epstein, R. (eds) *Handbook of Operations Research in Natural Resources*. Berlin: Springer.

Asche, F., Hansen, H., Tveterås, R. & Tveterås, S. (2009a) The salmon disease crisis in Chile. *Marine Resource Economics* **24**, 405–411.

Asche, F., Roll, K.H. & Tveterås, R. (2009b) Economic inefficiency and environmental impact: an application to aquaculture production. *Journal of Environmental Economics and Management* **58**, 93–105.

Asche, F., Roll, K.H. & Trollvik, T. (2009c) New aquaculture species: the white-fish market. *Aquaculture Economics and Management* **13**, 76–93.

Austreng, E. (1994) *Historical Development of Salmon Feed*. Annual Report 1993, Institute for Aquaculture Research (AKVAFORSK), Ås, Norway.

Bell, G., Torstensen, B. & Sargent, J. (2005) Replacement of marine fish oils with vegetable oils in feeds for farmed salmon. *Lipid Technology* **17**, 7–11.

Beveridge, M. (2004) *Cage Aquaculture*. Oxford: Blackwell Publishing Ltd.

Bird, P. (1986) Econometric estimation of world salmon demand. *Marine Resource Economics* **3**, 169–182.

Bjørndal, T. (1988) The optimal management of North Sea herring. *Journal of Environmental Economics and Management* **15**, 9–29.

Bjørndal, T. (1990) *The Economics of Salmon Aquaculture*. Oxford: Blackwell Publishing Ltd.

Bjørndal, T. (2002) The competitiveness of the Chilean salmon aquaculture industry. *Aquaculture Economics and Management* **6**, 97–116.

Bjørndal, T. & Aarland, K. (1999) Salmon aquaculture in Chile. *Aquaculture Economics and Management* **3**, 238–253.

Bjørndal, T. & Munro, G.R. (1998) The economics of fisheries management: a survey. In: Tietenberg, T. & Folmer, H. (eds) *The International Yearbook of Environmental and Resource Economics 1998/1999*. Cheltenham: Elgar.

Bjørndal, T. & Nøstbakken, L. (2003) Supply functions for North Sea herring. *Marine Resource Economics* **18**, 345–361.

Bjørndal, T. & Salvanes, K.G. (1995) Gains from deregulation? An empirical test for efficiency gains in the Norwegian fish farming industry. *Journal of Agricultural Economics* **46**, 113–126.

Bjørndal, T., Salvanes, K.G. & Andreassen, J.H. (1992) The demand for salmon in France: the effects of marketing and structural change. *Applied Economics* **24**, 1027–1034.

Bjørndal, T., Gordon, D.V. & Salvanes, K.G. (1994) Elasticity estimates of farmed salmon demand in Spain and Italy. *Empirical Economics* **4**, 419–428.

Bjørndal, T., Knapp, G.A. & Lem, A. (2003) *Salmon: A Study of Global Supply and Demand*. Rome: FAO/GLOBEFISH.

Black, E., Gowan, R., Rosenthal, H., Roth, E., Stechey, D. & Taylor, F.J.R. (1996) The cost of eutrophication from salmon farming: implications for policy. A comment. *Journal of Environmental Management* **49**, 105–109.

Cacho, O.J. (1997) Systems modeling and bioeconomic modeling in aquaculture. *Aquaculture Economics and Management* **1**, 45–64.

Clayton, P.L. & Gordon, D.V. (1999) From Atlantic to Pacific: price links in the US wild and farmed salmon market. *Aquaculture Economics and Management* **3**, 93–104.

Copes, P. (1972) Factor rents, sole ownership and the optimum level of fisheries exploitation. *Manchester School* **40**, 145–63.

Deaton, A.S. & Muellbauer, J. (1980) *Economics and Consumer Behavior*. New York: Cambridge University Press.

DeVoretz, D. (1982) An econometric demand model for Canadian salmon. *Canadian Journal of Agricultural Economics* **30**, 49–60.

DeVoretz, D.J. & Salvanes, K.G. (1993) Market structure for farmed salmon. *American Journal of Agricultural Economics* **75**, 227–233.

Eales, J. & Wessells, C.R. (1999) Testing separability of Japanese demand for meat and fish within differential demand systems. *Journal of Agricultural and Resource Economics* **24**, 114–126.

Eales, J., Durham, C. & Wessells, C.R. (1997) Generalized models of Japanese demand for fish. *American Journal of Agricultural Economics* **79**, 1153–1163.

Einen, O., Holmefjord, I., Åsgård, T. & Talbot, C. (1995) Auditing nutrient discharges from fish farms: theoretical and practical considerations. *Aquaculture Research* **26**, 701–713.

Engle, C.R. (2003) The evolution of farm management, production efficiencies, and current challenges to catfish production in the United States. *Aquaculture Economics and Management* **7**, 67–84.

Engle, C.R. & Valderrama, D. (2004) Economic effects of implementing selected components of best management practices (BMPs) for semi-intensive shrimp farms in Honduras. *Aquaculture Economics and Management* **8**, 157–177.

Ezekiel, M. (1938) The cobweb theorem. *Quarterly Journal of Economics* **52**, 255–280.

FAO (2009) *The State of World Fisheries and Aquaculture 2008*. Rome: Food and Agriculture Organization of the United Nations.

Fofana, A. & Jaffry, S. (2008) Measuring buyer power (oligopsony) of UK salmon retailers. *Marine Resource Economics* **23**, 485–506.

Folke, C., Kautsky, N. & Troell, M. (1994) The costs of eutrophication from salmon farming: implications for policy. *Journal of Environmental Management* **40**, 173–182.

Forsberg, O.I. & Guttormsen, A.G. (2005) A pigmentation model for farmed Atlantic salmon: non-linear regression analysis of published experimental data. *Aquaculture* **253**, 415–420.

Forsberg, O.I. & Guttormsen, A.G. (2006) Modeling optimal dietary pigmentation strategies in farmed Atlantic salmon: application of mixed-integer non-linear mathematical programming techniques. *Aquaculture* **261**, 118–124.

Gjedrem, T. (2000) Genetic improvement of cold-water fish species. *Aquaculture Research* **31**, 25–33.

Gjedrem, T. & Baranski, M. (2009) *Selective Breeding in Aquaculture: An Introduction*. New York: Springer.

Goldberg, P.K. & Knetter, M.M. (1997) Goods prices and exchange rates: what have we learned? *Journal of Economic Literature* **35**, 1243–1272.

Gordon, D.V., Salvanes, K.G. & Atkins, F. (1993) A fish is a fish is a fish: testing for market linkage on the Paris fish market. *Marine Resource Economics* **8**, 331–343.

Gordon, D.V., Bjørndal, T., Dey, M. & Talukder, R.K. (2008) An intra-farm study of production factors and productivity for shrimp farms in Bangladesh: an index approach. *Marine Resource Economics* **23**, 411–424.

Grainger, R.J.R. & Garcia, S.M. (1996) *Chronicles of Marine Fishery Landings (1950–1994): Trend Analysis and Fisheries Potential*. Rome: FAO Fisheries Technical Paper No. 359.

Grimnes, A., Finstad, B., Bjørn, P.A., Tovslid, B.M. & Lund, R. (1998) Registration of sea lice on salmon, salmon trout and chars in 1997 (In Norwegian: Registrering av lakselus på laks, sjøørret og sjørøye i 1997). NINA Oppdragsmelding 525. Oslo.

Guillotreau, P., Le Grel, L. & Simioni, M. (2005) Price-cost margins and structural change: sub-contracting within the salmon marketing chain. *Review of Development Economics* **9**, 581–597.

Guttormsen, A.G. (1999) Forecasting weekly salmon prices: risk management in salmon farming. *Aquaculture Economics and Management* **3**, 159–166.

Guttormsen, A.G. (2002) Input factor substitutability in salmon aquaculture. *Marine Resource Economics* **17**, 91–102.

Guttormsen, A.G. (2008) Faustman in the sea: optimal harvesting of farmed fish. *Marine Resource Economics* **23**, 401–410.

Hannesson, R. (2003) Aquaculture and fisheries. *Marine Policy* **27**, 169–178.

Heaps, T. (1993) The optimal feeding of farmed fish. *Marine Resource Economics* **8**, 88–99.

Heaps, T. (1995) Density dependent growth and the culling of farmed fish. *Marine Resource Economics* **10**, 285–298.

Heen, K., Mohnahan, R.L. & Utter, F. (1993) *Salmon Aquaculture*. Oxford: Blackwell Publishing Ltd.

Hempel, E. (1997) *The Market for Pelagics*. Trondheim: Hempel Consult.

Herrmann, M.L. & Lin, B.H. (1988) The demand and supply of Norwegian Atlantic salmon in the United States and the European Community. *Canadian Journal of Agricultural Economics* **38**, 459–471.

Herrmann, M., Mittelhammer, R.C. & Lin, B.H. (1992) Applying Almon-type polynomials in modelling seasonality of the Japanese demand for salmon. *Marine Resource Economics* **7**, 3–13.

Herrmann, M.L., Mittelhammer, R.C. & Lin, B.H. (1993) Import demand for Norwegian farmed Atlantic salmon and wild Pacific salmon in North America, Japan and the EC. *Canadian Journal of Agricultural Economics* **41**, 111–125.

Holland, D. & Wessells, C.R. (1998) Predicting consumer preferences for fresh salmon: the influence of safety inspection and production method attributes. *Agricultural and Resource Economics Review* **27**, 1–14.

Holmer, M., Black, K., Duarte, C.M., Marba, N. & Karakassis, I. (2008) *Aquaculture in the Ecosystem*. Berlin: Springer.

Iwama, G.K. & Tautz, A.F. (1981) A simple growth model for salmonids in hatcheries. *Canadian Journal of Fisheries and Aquatic Sciences* **38**, 649–56.

Jaffry, S., Pascoe, S., Taylor, G. & Zabala, U. (2000) Price interactions between salmon and wild caught fish species on the Spanish market. *Aquaculture Economics and Management* **4**, 157–168.

Jaffry, S., Fofana, A. & Murray, A.D. (2003) Testing for market power in the UK salmon retail sector. *Aquaculture Economics and Management* **7**, 293–308.

Johnson, A., Durham, C.A. & Wessells, C.R. (1998) Seasonality in Japanese household demand for meat and seafood. *Agribusiness* **14**, 337–351.

Johnston, R.J. & Roheim, C.A. (2006) A battle of taste and environmental convictions for ecolabeled seafood: a contingent ranking experiment. *Journal of Agricultural and Resource Economics* **31**, 283–300.

Johnston, R.J., Wessells, C.R., Donath, H. & Asche, F. (2001) Measuring consumer preferences for ecolabeled seafood: an international comparison. *Journal of Agricultural and Resource Economics* **26**, 20–39.

Kabir, M. & Ridler, N.B. (1984) The demand for Atlantic salmon in Canada. *Canadian Journal of Agricultural Economics* **32**, 560–568.

Karagiannis, G. & Katranidis, S.D. (2000) A production function analysis of seabass and seabream production in Greece. *Journal of the World Aquaculture Society* **31**, 297–305.

Karp, L., Sadeh, A. & Griffin, W.L. (1986) Cycles in agricultural production. The case of aquaculture. *American Journal of Agricultural Economics* **68**, 553–561.

Katranidis, S.D., Tzouvelekas, V. & Karagiannis, G. (2002) Measuring and attributing technical inefficiencies of seabass and seabream production in Greece. *Applied Economics Letters* **9**, 519–522.

Kinnucan, H. (1995) Catfish aquaculture in the United States: five propositions about industry growth and policy. *World Aquaculture* **26**, 13–20.

Kinnucan, H.W. & Myrland, Ø. (2000) Optimal advertising levies with application to the Norway–EU salmon agreement. *European Review of Agricultural Economics* **27**, 39–57.

Kinnucan, H.W. & Myrland, Ø. (2001) A note on measuring returns to nonprice export promotion. *Agribusiness* **17**, 423–434.

Kinnucan, H. & Myrland, Ø. (2002a) Seasonal allocation of an advertising budget. *Marine Resource Economics* **17**, 103–120.

Kinnucan, H.W. & Myrland, Ø. (2002b) The relative impact of the Norway–EU salmon agreement: a mid-term assessment. *Journal of Agricultural Economics* **53**, 195–220.

Kinnucan, H.W. & Myrland, Ø. (2003) Free-rider effects of generic advertising: the case of salmon. *Agribusiness: An International Journal* **19**, 315–324.

Kinnucan, H. & Myrland, Ø. (2005) Effects of income growth and tariffs on the world salmon market. *Applied Economics* **37**, 1967–1978.

Kinnucan, H.W. & Myrland, Ø. (2006) The effectiveness of antidumping measures: some evidence for farmed Atlantic salmon. *Journal of Agricultural Economics* **57**, 459–477.

Kinnucan, H.W. & Myrland, Ø. (2007) On generic vs. brand promotion of farm products in foreign markets. *Applied Economics* **40**, 673–684.

Kinnucan, H. & Wessells, C.R. (1997) Marketing research paradigms for aquaculture. *Aquaculture Economics and Management* **1**, 73–86.

Kinnucan, H.W., Asche, F., Myrland, Ø. & Roheim, C.A. (2003) Advances in economics of marketing and implications for aquaculture development. *Aquaculture Economics and Management* **7**, 35–53.

Knapp, G., Roheim, C.A. & Anderson, J.L. (2007) *The Great Salmon Run: Competition Between Wild and Farmed Salmon*. Washington, DC: TRAFFIC.

Kolstad, K., Grisdale-Helland, B. & Gjerde, B. (2004) Family differences in feed efficiency in Atlantic salmon (*Salmo salar*). *Aquaculture* **241**, 169–177.

Kristofersson, D. & Anderson, J.L. (2006) Is there a relationship between fisheries and farming? Interdependence of fisheries, animal production and aquaculture. *Marine Policy* **30**, 721–725.

Krugman, P. & Obstfeldt, M. (1994) *International Economics: Theory and Policy*. New York: Harper-Collins.

Kumbhakar, S.C. (2001) Estimation of profit functions when profit is not maximum. *American Journal of Agricultural Economics* **83**, 1–19.

Kumbhakar, S.C. (2002a) Risk preferences and technology: a joint analysis. *Marine Resource Economics* **17**, 77–89.

Kumbhakar, S.C. (2002b) Specification and estimation of production risk, risk preferences and technical efficiency. *American Journal of Agricultural Economics* **84**, 8–22.

Kumbhakar, S.C. & Tveterås, R. (2003) Risk preferences, production risk and firm heterogeneity. *Scandinavian Journal of Economics* **105**, 275–293.

Kvaløy, O. & Tveterås, R. (2008) Cost structure and vertical integration between farming and processing. *Journal of Agrucultural Economics* **59**, 296–311.

Larsen T.A. & Kinnucan H.W. (2009) The effect of exchange rates on international marketing margins. *Aquaculture Economics and Management* **13**, 124–137.

Leikang, O.-I. (2007) *Aquaculture Engineering*. Oxford: Blackwell Publishing Ltd.

Leung, P.S. & C. Shang, C. (1989) Modeling prawn production management system. A dynamic Markov decision approach. *Agricultural Systems* **29**, 5–20.

Leung, P.S., Lee, C.-H. & O'Bryen, P.J. (2007) *Species and System Selection for Sustainable Aquaculture*. Ames, IA: Blackwell Publishing Ltd.

Martin, S. (1993) *Advanced Industrial Economics*. Oxford: Blackwell Publishing Ltd.

Mistiaen, J.A. & Strand, I. (1998) Optimal feeding and harvest time for fish with weight-dependent prices. *Marine Resources Economics* **13**, 231–246.

Moksness, E., Kjørsvik, E. & Olsen, Y. (2003) *Culture of Cold-water Marine Fish*. Oxford: Blackwell Publishing Ltd.

Mozaffarian, D. & Rimm, E.B. (2006) Fish intake, contaminants, and human health: evaluating the risks and the benefits. *Journal of the American Medical Association* **296**, 1885–1899.

Munro, G.R. & Scott, A.D. (1985) The economics of fisheries management. In: Kneese, A.V. & Sweeny, J.L. (eds) *Handbook of Natural Resource and Energy Economics*. Amsterdam: North Holland.

Murray, A.D. & Fofana, A. (2002) The changing nature of UK fish retailing. *Marine Resource Economics* **17**, 335–340.

Nakomoto, A. (2000) *The Japanese Seafood Market*. Bergen: Institute for Research in Economics and Business Administration.

Naylor, R.L., Goldburg, R.J., Primavera, J. *et al.* (2000) Effects of aquaculture on world fish supplies. *Nature* **405**, 1017–1024.

Nielsen, M., Setala, J., Laitinen, J., Saarni, K., Virtanen, J. & Honkanen, A. (2007) Market integration of farmed trout in Germany. *Marine Resource Economics* **22**, 195–213.

Nielsen, M., Smit, J. & Guillen, J. (2009) Market integration of fish in Europe. *Journal of Agricultural Economics* **60**, 367–385.

Nilsen, O.B. (2010) Learning-by-doing or technologcal leapfroging: production frontiers and efficiency measurement in Norwegian salmon aquaculture. *Aquaculture Economics and Management* **14**, 97–119.

Norman-López, A. (2009) Competition between different wild and farmed species: the US tilapia market. *Marine Resource Economics* **24**, 237–252.

Norman-López, A. & Asche, F. (2008) Competition between imported tilapia and US catfish in the US market. *Marine Resource Economics* **23**, 199–214.

Norman-López, A. & Bjørndal, T. (2009) Is tilapia the same product worldwide or are markets segmented? *Aquaculture Economics and Management* **13**, 138–154.

Oglend, A. & Sikveland, M. (2008) The behaviour of salmon price volatility. *Marine Resource Economics* **23**, 507–526.

Oglend, A. & Tveterås, R. (2009) Spatial diversification in Norwegian aquaculture. *Aquaculture Economics and Management* **13**, 94–111.

Olaussen, J.O. (2007) Playing chicken with salmon. *Marine Resource Economics* **22**, 173–193.

Olson, T.K. & Criddle, K.R. (2008) Industrial evolution: a case study of Chilean salmon aquaculture. *Aquaculture Economics and Management* **12**, 89–106.

Østby, S. (1999) A technical note on input price proxies used in salmon farming industry studies. *Marine Resource Economics* **14**, 215–223.

Prusa, T.J. (1996) The trade effects of US antidumping actions. In: Feenstra, R.C. (ed.) *Effects of U.S. Trade Proctection and Promotion Policies*. Chicago: University of Chicago Press.

Rizzo, G. & Spagnolo, M. (1996) A model for the optimal management of sea bass *Dicentrarchus labrax* aquaculture. *Marine Resource Economics* **11**, 267–286.

Roheim, C.A. (2003) Early indications of market implications from the Marine Stewardship Council's ecolabeling of seafood. *Marine Resource Economics* **16**, 95–104.

Roheim, C.A. (2009) An evaluation of sustainable seafood guides: implications for environmental groups and the seafood industry. *Marine Resource Economics* **24**, 301–310.

Salvanes, K.G. (1993) Public regulation and production factor misallocation: a restricted cost function approach for the Norwegian aquaculture industry. *Marine Resource Economics* **8**, 50–64.

Salvanes, K.G. & DeVoretz, D.J. (1997) Household demand for fish and meat products: separability and demographic effects. *Marine Resource Economics* **12**, 37–55.

Sharma, K.R. & Leung, R.S. (2003) A review of production frontier analysis for aquaculture management. *Aquaculture Economics and Management* **7**, 15–34.

Shaw, S.A. & Muir, J.F. (1987) *Salmon: Economics and Marketing*. London: Croom Helm.

Smith, M.D., Roheim, C.A., Crowder, L.B. *et al.* (2010) Sustainability and global seafood. *Science* **327**, 784–786.

Steen, F. & Salvanes, K.G. (1999) Testing for market power using a dynamic oligopoly model. *International Journal of Industrial Organization* **17**, 147–177.

Stigler, G.J. (1969) *The Theory of Price*. London: Macmillan.

Talbot, C. (1993) Some aspects of the biology of feeding and growth in fish. *Proceedings of the Nutrition Society* **52**, 403–416.

Thodesen, J., Grisdale-Helland, B., Helland, S.J. & Gjerde, B. (1999) Feed intake, growth and feed utilization of offspring from wild and selected Atlantic salmon *Salmo salar*. *Aquaculture* **180**, 237–246.

Thodesen, J., Gjerde, B., Grisdale-Helland, B. & Storebakken, T. (2001) Genetic variation in feed intake, growth and feed utilization in Atlantic salmon *Salmo salar*. *Aquaculture* **194**, 273–281.

Torstensen, B.E., Bell, J.G., Rosenlund, G. *et al.* (2005) Tailoring of Atlantic salmon (*Salmo salar* L.) flesh lipid composition and sensory quality by replacing fish oil with a vegetable oil blend. *Journal of Agricultural Food Chemistry* **53**, 10166–10178.

Tully, O. & Nolan, D.T. (2002) A review of the population biology and host–parasite interactions of the sea louse *Lepeophtheirus salmonis* (Copepoda: Caligidae). *Parasitology* **124**, S165–S182.

Turcini, G.M., Torstensen, B. & Ng, W.-K. (2009) Fishoil replacement in finfish nutrition. *Reviews in Aquaculture* **1**, 10–57.

Tveterås, R. (1999) Production risk and productivity growth: some findings for Norwegian salmon aquaculture. *Journal of Productivity Analysis* **12**, 161–179.

Tveterås, R. (2000) Flexible panel data models for risky production technologies with an application to salmon aquaculture. *Econometric Reviews* **19**, 367–389.

Tveterås, R. (2002) Industrial agglomeration and production costs in Norwegian aquaculture. *Marine Resource Economics* **17**, 1–22.

Tveterås, R. & Batteese, G.M. (2006) Agglomeration externalities, productivity and technical inefficiency. *Journal of Regional Science* **46**, 605–625.

Tveterås, R. & Heshmati, A. (2002) Patterns of productivity growth in the Norwegian salmon farming industry. *International Review of Economics and Business* **49**, 367–393.

Tveterås, S. (2002) Norwegian salmon aquaculture and sustainability: the relationship between environmental quality and industry growth. *Marine Resource Economics* **17**, 121–132.

Valderrama, D. & Anderson, J.L. (2010) Market interactions between aquaculture and common-property fisheries: recent evidence from the Bristol Bay sockeye salmon fishery in Alaska. *Journal of Environmental Economics and Management* **59**, 115–128.

Varian, H.R. (1992) *Microeconomic Analysis*. New York: Norton.

Virtanen, J., Setala, J., Saarni, K. & Honkanen, A. (2005) Finnish salmon trout: discriminated in the European market. *Marine Resource Economics* **20**, 113–119.

Vukina, T. & Anderson, J.L. (1993) A state-space forecasting approach to optimal intertemporal cross-hedging. *American Journal of Agricultural Economics* **75**, 416–424.

Vukina, T. & Anderson, J.L. (1994) Price forecasting with state-space models of nonstationary time series: the case of the Japanese salmon market. *Computers and Mathematics with Applications* **27**, 45–62.

Wallace, J. (1993) Environmental considerations. In: Heen, K., Mohnahan, R.L. & Utter, F. (eds) *Salmon Aquaculture*. Oxford: Blackwell Publishing Ltd.

Wessells, C.R. (2002) Markets for seafood attributes. *Marine Resource Economics* **17**, 153–162.

Wessells, C.R. & Anderson, J.L. (1992) Innovations and progress in seafood demand and market analysis. *Marine Resource Economics* **7**, 209–288.

Wessells, C.R. & Wilen, J.E. (1993a) Economic analysis of Japanese household demand for salmon. *Journal of the World Aquaculture Society* **24**, 361–378.

Wessells, C.R. & Wilen, J.E. (1993b) Inventory dissipation in the Japanese wholesale salmon market. *Marine Resource Economics* **8**, 1–16.

Wessells, C.R. & Wilen, J.E. (1994) Seasonal patterns and regional preferences in Japanese household demand for seafood. *Canadian Journal of Agricultural Economics* **42**, 87–103.

Wessells, C.R., Johnston, R.J. & Donath, H. (1999) Assesing consumer preferences for ecolabeled seafood: the influence of species, certifier and household attributes. *American Journal of Agricultural Economics* **81**, 1084–1089.

Willett, W.C. (2005) Fish: balancing health risks and benefits. *American Journal of Preventive Medicine* **29**, 321–322.

Xie, J., Kinnucan, H.W. & Myrland, Ø. (2008) The effects of exchange rates on export prices of farmed salmon. *Marine Resource Economics* **23**, 439–57.

Young, J.A. & Muir, J. (2002) Tilapia: both fish and fowl? *Marine Resource Economics* **17**, 163–173.

Index

advertising, 4, 21, 35, 133–5
Africa, 2, 16, 158
agglomeration, 62
airfreight, 4, 95, 110, 125, 126
America, 4, 24, 26, 61, 66, 98, 123, 124,
    126, 132, 152, 158
anchoveta, 66, 68
animal husbandry, 7
antibiotics, 52, 54, 74, 76–8
Aquachile, 40
aquaculture, 1–5, 7–12, 15–17, 20,
    22–4, 27, 34, 38–43, 53–6, 63, 65, 66,
    69–74, 76–8, 80, 81, 83, 101, 119,
    123, 124, 129, 131, 135, 136, 142,
    146, 148, 149, 151, 152, 154, 156–8,
    160–164, 167, 185, 201, 205, 208,
    211, 214, 217
Asia, 24, 61, 83, 123, 134, 152, 158, 160
astaxanthin, 52, 53, 63
Atlantic, 9, 152
Atlantic salmon, 1, 2, 10, 11, 17–19,
    23–6, 29–32, 55, 56, 59–62, 85–8,
    90–96, 98–100, 103, 106, 107, 124–8,
    134, 139, 140, 187
Australia, 30, 31, 62, 160
autumn, 10–13, 21, 95, 97, 98, 114,
    139, 205

bioeconomic, 145, 165, 172
biological, 5, 7–10, 12, 16, 30, 38, 56,
    68, 74, 123, 125, 145, 160, 162–5,
    170–172, 184, 187, 190, 195, 200
biology, 16, 114
biomass, 8, 14, 36, 54, 68, 74, 164–7,
    169, 172–4, 181–3, 187, 189–94,
    197–9, 201, 202, 204

biophysical, 13, 58, 171
brand, 87, 133, 134
Brazil, 24, 26, 83, 152, 157
breeding, 7, 10, 11, 49, 55, 56, 64, 78,
    158, 170
broodstock, 10, 20, 35, 79
buyer, 3, 24, 65, 69, 72, 107–9, 111,
    120–122, 124, 128, 134, 142

cage, 15, 74, 75, 77, 79
Canada, 17, 18, 29, 32, 34, 35, 40, 41,
    50, 54, 59–62, 84, 90, 96, 98–100,
    103–5, 122, 132, 136, 148
capacity, 10, 15, 21, 22, 34, 35, 38, 44,
    74, 124, 131, 145, 170, 179, 189,
    200–202, 204, 215, 219
capital, 4, 12, 24, 26, 35, 48–50, 61, 63,
    156, 167, 174, 175, 177, 188, 201,
    204, 205, 207–13, 215, 217, 218
capital cost, 156, 209–11
cash-flow, 165, 169, 181, 187–90,
    197–9, 208, 211–14, 217–19
catfish, 41, 100, 105, 162
Cermaq, 40, 41
chemicals, 74, 76–8
cherry, 9, 17, 18, 23
Chile, 3, 9, 11, 12, 17, 18, 20, 23–7, 30,
    31, 34, 35, 38–40, 41, 42, 45, 50, 54,
    59–64, 66–8, 83, 86, 89, 90, 96, 98,
    100, 103–7, 112, 117, 120, 122, 123,
    125–8, 132, 139–41, 199, 205, 217
China, 8, 67, 103, 152, 157, 158
Chinese, 8
chinook, 9, 11, 17, 18, 23, 29, 31–3, 96,
    101, 124, 126
chum, 9, 31–3, 94–8, 101, 124

The Economics of Salmon Aquaculture, Second Edition. Frank Asche and Trond Bjørndal.
© 2011 Frank Asche and Trond Bjørndal. Published 2011 by Blackwell Publishing Ltd.